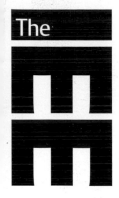

Guidance Note 6

Protection Against Overcurrent

17th IEE Wiring Regulations Seventeenth Edition
BS 7671:2008 R~~equirements~~ ...ations

Published by The Institution of Engineering and Technology, London, United Kingdom

The Institution of Engineering and Technology is registered as a Charity in England & Wales (no. 211014) and Scotland (no. SCO38698).

The Institution of Engineering and Technology is the new institution formed by the joining together of the IEE (The Institution of Electrical Engineers) and the IIE (The Institution of Incorporated Engineers). The new Institution is the inheritor of the IEE brand and all its products and services, such as this one, which we hope you will find useful. The IEE is a registered trademark of the Institution of Engineering and Technology.

First published 1993 (0 85296 540 0)
Reprinted (with minor amendments) 1993
Second edition (incorporating Amendment No. 1 to BS 7671:1992) 1996 (0 85296 871 X)
Third edition (incorporating Amendment No. 2 to BS 7671:1992) 1999 (0 85296 959 7)
Fourth edition (incorporating Amendment No. 1 to BS 7671:2001) 2003 (0 85296 994 5)
Reprinted (incorporating Amendment No. 2 to BS 7671:2001) 2004
Fifth edition (incorporating BS 7671:2008) 2009 (978-0-86341-860-0)

Copies of this publication may be obtained from:
The Institution of Engineering and Technology
PO Box 96
Stevenage
SG1 2SD, UK
Tel: +44 (0)1438 767328
Email: sales@theiet.org
www.theiet.org/publishing/books/wir-reg/

ISBN 978-0-86341-860-0

Typeset in the UK by Carnegie Book Production, Lancaster
Printed in the UK by Printwright Ltd, Ipswich

Contents

Chapter 5 Equations for the calculation of short-circuit current 57

Chapter 6 Equations for the calculation of earth fault current 67

Chapter 7 Selection of conductor size 77

Appendix A Calculation of reactance 83

Appendix B Calculation of k for other temperatures 89

Index 91

Cooperating organisations

The Institution of Engineering and Technology acknowledges the contribution made by the following organisations in the preparation of this Guidance Note.

British Cables Association
J.M.R. Hagger BTech(Hons) AMIMMM
C.K. Reed IEng MIET

British Electrotechnical & Allied Manufacturers Association Ltd
P.D. Galbraith IEng MIET

BEAMA Installation Ltd
P. Sayer IEng MIET GCGI
Eur Ing M.H. Mullins BA CEng FIEE FIIE

City & Guilds of London Institute
H.R. Lovegrove IEng FIIE

Electrical Contractors' Association
D. Locke BEng(Hons) IEng MIET ACIBSE

ERA Technology Ltd
M.W. Coates BEng

Health and Safety Executive
K.J. Morton BSc CEng MIEE

Institution of Engineering and Technology
M. Coles BEng(Hons) MIET
G.D. Cronshaw IEng FIET (Editor)
P.E. Donnachie BSc CEng FIET
J.F. Elliott BSc(Hons) IEng MIET

Lighting Association Ltd
L. Barling

SELECT (Electrical Contractors' Association of Scotland)
D. Millar IEng MIET MILE

Acknowledgements

References to British Standards, CENELEC Harmonization Documents and International Electrotechnical Commission standards are made with the kind permission of BSI. Complete copies can be obtained by post from:

BSI Customer Services
389 Chiswick High Road
London W4 4AL
Tel: +44 (0)20 8996 9001
Fax: +44 (0)20 8996 7001
Email: orders@bsi-global.com

BSI also maintains stocks of international and foreign standards, with many English translations. Up-to-date information on BSI standards can be obtained from the BSI website: www.bsi-global.com

Most of the illustrations within this publication were provided by Rod Farquhar Design: www.rodfarquhar.co.uk

Cover design and illustration were created by The Page Design: www.thepagedesign. co.uk

It is strongly recommended that anyone involved in work on or near electrical installations possesses a copy of the *Memorandum of guidance on the Electricity at Work Regulations 1989* (HSR25) published by the Health and Safety Executive.

Copies of Health and Safety Executive documents and approved codes of practice (ACOP) can be obtained from:

HSE Books
PO Box 1999
Sudbury, Suffolk CO10 2WA
Tel: +44 (0)1787 881165
Email: hsebooks@prolog.uk.com
Web: www.hsebooks.com

The HSE website is www.hse.gov.uk

Preface

This Guidance Note is part of a series issued by the Institution of Engineering and Technology to explain and enlarge upon the requirements in BS 7671:2008, the 17th Edition of the IEE Wiring Regulations.

Note that this Guidance Note does not ensure compliance with BS 7671. It is intended to explain some of the requirements of BS 7671, but readers should always consult BS 7671 to satisfy themselves of compliance.

The scope generally follows that of BS 7671; the relevant Regulations and Appendices are noted in the margin. Some Guidance Notes also contain material not included in BS 7671:2008 but which was included in earlier editions of the Wiring Regulations. All of the Guidance Notes contain references to other relevant sources of information.

Electrical installations in the United Kingdom that comply with BS 7671 are likely to satisfy Statutory Regulations such as the Electricity at Work Regulations 1989, however this cannot be guaranteed. It is stressed that it is essential to establish which Statutory and other Regulations apply and to install accordingly. For example, an installation in premises subject to licensing may have requirements different from, or additional to, BS 7671 and these will take precedence.

Introduction

This Guidance Note is concerned primarily with Chapter 43 of BS 7671 – Protection against overcurrent – and has been revised to align with BS 7671:2008 *IEE Wiring Regulations 17th Edition*.

Chapter 43 addresses the protection of live circuit conductors, i.e. line and neutral conductors, from the effects of overcurrent. Many of the detailed requirements remain essentially unchanged; however, the Chapter 43 basic requirements and Section 473 application requirements of the 16th Edition have been merged in the 17th Edition, resulting in a modified Chapter 43.

The requirements for positioning of overload devices and also for their omission remain more or less unchanged, except that regarding omission of overload protection, two additional examples are included. These are a circuit supplying a safety service, such as a fire alarm or gas alarm, and a circuit supplying medical equipment used for life support.

Appendix 4 of BS 7671 has been completely revised. The term 'rating factor' replaces the term 'correction factor' throughout the tables. New installation methods have been added for flat twin and earth cables in thermal insulation. Rating factors have been added for cables buried direct in the ground. Additional rating factors for cables in free air have been added, resulting in an increased number of installation methods and reference methods.

In addition, BS 7671:2008 includes a new, informative Appendix 11 dealing with the effects of harmonic currents on balanced three-phase systems and also an informative Appendix 10 dealing with protection of conductors in parallel against overcurrent.

Finally, BS 7671:2008 includes a new Appendix 15, which sets options for the design of ring and radial circuits for household and similar premises in accordance with Regulation 433.1, using socket-outlets and fused connection units.

Neither BS 7671 nor the Guidance Notes are design guides. It is essential to prepare a full design and specification prior to commencement or alteration of an electrical installation.

The design and specification should set out the requirements and provide sufficient information to enable competent persons to carry out the installation and to commission it. The specification must include a description of how the system is to operate and all the design and operational parameters. It must provide for all the commissioning procedures that will be required and for the provision of adequate information to the user. This should be by means of an operation and maintenance manual or schedule, 514.9 and 'as fitted' drawings if necessary.

It must be noted that it is a matter of contract as to which person or organisation is responsible for the production of the parts of the design, specification, construction and verification of the installation and any operational information.

The persons or organisations who may be concerned in the preparation of the works include:

> The Designer
> The CDM Coordinator
> The Installer
> The Supplier of Electricity (Distributor)
> The Installation Owner (Client) and/or User
> The Architect
> The Fire Prevention Officer
> All Regulatory Authorities
> Any Licensing Authority
> The Health and Safety Executive

132.1 In producing the design, advice should be sought from the installation owner and/or user as to the intended use. Often, as in a speculative building, the intended use is unknown. The specification and/or the operational manual must set out the basis of use for which the installation is suitable.

133.1
Section 511 Precise details of each item of equipment should be obtained from the manufacturer and/or supplier and compliance with appropriate standards confirmed.

The operational manual must include a description of how the system as installed is to operate and all commissioning records. The manual should also include manufacturers' technical data for all items of switchgear, luminaires, accessories, etc. and any special instructions that may be needed.

The Health and Safety at Work etc. Act 1974 Section 6 and the Construction (Design and Management) Regulations 2007 are concerned with the provision of information, and guidance on the preparation of technical manuals is given in the BS 4884 series *Technical manuals* and the BS 4940 series *Technical information on constructional products and services*. The size and complexity of the installation will dictate the nature and extent of the manual.

The regulations concerning protection against overcurrent

1

1.1 Scope

1.1.1 Fundamental principles
The fundamental principles of Chapter 13 of BS 7671 include two regulations concerning overcurrent.

Chap 13

The first of these requires persons or livestock to be protected against injury, and property to be protected against damage, caused by excessive temperatures or electromechanical stresses due to any overcurrents likely to arise in live conductors.

131.4

It is worth noting that the words 'so far as is reasonably practicable' no longer appear in the fundamental regulation in the 17th Edition.

The second regulation requires conductors other than live conductors, and any other parts intended to carry a fault current, to be capable of carrying that current without attaining an excessive temperature. Electrical equipment, including conductors, must be provided with mechanical protection against electromechanical stresses of fault currents as necessary to prevent injury or damage to persons, including livestock or property.

131.5

1.1.2 Protective measures
Measures to provide protection against overcurrents are contained in Chapter 43 and in a number of chapters and sections of Part 5.

Chap 43
Part 5

Chapter 43 gives the main requirements for the protection of live conductors against the effects of both overload current and fault current.

Sections 533 and 536 contain regulations for the selection of overcurrent protective devices.

1.1.3 Load equipment and flexible cords
It must be remembered that the Regulations deal primarily with the fixed equipment of an installation. Protection of conductors in accordance with Chapter 43 does not necessarily protect the equipment connected to an installation nor the flexible cables and cords connecting such equipment to the fixed part of an installation. This subject is of importance where BS industrial-type plugs and sockets are to be used. It may be necessary to consider the feasibility of a limitation on the current rating of such circuits, to introduce local protection, or to require minimum conductor sizes for the flexible cords.

1.2 Nature of overcurrent and protection

1.2.1 General

430.1 Every circuit must be protected by one or more devices which automatically interrupt the supply in the event of overcurrent.

132.8 The object of the Regulations is to ensure that any overcurrent does not persist
430.3 long enough to cause damage to equipment/property or risk of injury to persons or livestock.

Part 2 An *overcurrent* is any current which exceeds the current-carrying capacity of the circuit conductors.

For overload current, this is determined by the rated current or current setting of the overcurrent protective device for that circuit, and the current-carrying capacity of its
433.1.1 conductors, by the application of Regulation 433.1.1. The conductors must also be protected against fault currents.

1.2.2 Overload current and fault current

Part 2 The term overcurrent includes both *overload current* and *fault current*.

Part 2 Fault current may be either *short-circuit current* or *earth fault current*.

The difference, in principle, between these lies in the reason for their occurrence, not in their magnitudes. An overload is imposed on a sound circuit by an abnormal situation at the load, whereas fault current is due to an insulation failure or bridging of insulation in the circuit or its terminations.

In practice, however, the difference between the two types of overcurrent tends to be one of current magnitude and duration, because this makes for a convenient
Appx 3 classification of the protective devices required. The total time/current characteristic of a protective device, or combination of devices, should provide protection for any combination of current and time which could result in overheating or mechanical overstressing. This total characteristic is exemplified in the single characteristic of a fuse or the characteristic of a circuit-breaker, either thermal/magnetic or electronic.

It is generally sufficient to look at the two extremes of this characteristic: the high current short duration part which, because it usually protects from faults in the circuit itself, is referred to as fault protection; and the long duration lower current part, usually protecting against excessive load, which is referred to as overload protection.

Furthermore, experience shows that these two extremes cover practically all abnormal currents and where calculation of the values of current is the most practicable.

It is accepted that actions such as direct-on-line motor starting, or capacitor switching, could result in currents of a magnitude similar to some fault currents, while a high resistance earth fault or a fault in equipment could result in a current of the same magnitude as those arising from excessive load.

1.2.3 General protection characteristic

The total characteristic of circuit protection complying with the Regulations is intended to be such that if the protection is adequate at the two extremes as described above, it will be satisfactory for any time/current combination in between.

However, for practical reasons it is sometimes not feasible to provide complete protection for very small overloads and attention is drawn to the comments in paragraph 2.2.4 on Regulation 433.1 with regard to load assessment.

433.1

1.3 Statutory requirements

Regulation 11 of the Electricity at Work Regulations 1989 requires that efficient means, suitably located, shall be provided to protect against excess current in every part of a system, as may be necessary to prevent danger.

It is important that the designer and the installer consult the relevant documents listed in Appendix 2 of BS 7671, together with those statutory regulations and memoranda which apply to the particular installation concerned.

Appx 2

Regulation 114.1 sets out the relationship of BS 7671:2008 with statutory regulations, including the Electricity Safety, Quality and Continuity Regulations 2002. The statutory regulations are also listed in Appendix 2 of BS 7671:2008.

114.1

There may also be local regulations which apply to installations in certain premises, such as where there is public access, certain types of entertainment, housing for the aged or infirm, etc.

The impact of any Regulation which may bear on the function or cost of an installation should be brought to the attention of the customer at the earliest opportunity.

1.4 Omission of protection

Protection against either overload or fault current may be omitted where unexpected disconnection could cause danger. The designer should consult Sections 433 and 434 to establish the conditions under which protection may be omitted.

433.3.3
434.3

In addition, Regulation 433.3.1 sets out the circumstances where a device for protection against overload need not be provided, as described in detail in section 2.4 of this Guidance Note.

433.3.1

The omission of overload or fault current protection, or a reduction in its effectiveness, can only be justified if danger is prevented by other means, or where the opening of the circuit would cause greater danger than the overload or fault condition.

This calls for a careful consideration of the balance of risks involved. The integrity of the equipment involved, including its ability to contain safely any possible effects of an overcurrent, such as arcs, hot particles, fire and dangerous fumes etc., should be examined.

The design should mitigate the effects of the omission of the overcurrent device, for example by rating the circuit for any current reasonably likely to arise.

1.5 Protective devices

432.1 Protection may be provided by a single device which detects and interrupts both overload and fault current. Alternatively, separate devices may be used for each task.

There are special provisions for protection of motor starter circuits; see section 1.7.3.

1.6 Duration of overcurrent

While BS 7671 does not impose a specific upper limit on the duration of overcurrent, except where there is a risk of shock or of damage to equipment, it is prudent to select circuit protection so that fault current does not persist any longer than is absolutely necessary. In a practical installation there is little or no control over the paths fault current may take, especially earth fault current, and danger or damage may occur in unexpected ways and to items not directly associated with the circuit.

434.5.2
Table 43.1
Note that the data provided in Regulation 434.5.2 (Table 43.1) are valid for fault disconnection times not exceeding 5 seconds, and this tends to set an implied upper limit to fault current duration.

The duration of small overloads should be controlled by suitable attention to load assessment; see the guidance on Regulation 433.1 in section 2.2.4.

1.7 Coordination and discrimination

Section 536 Section 536 'Coordination of protective devices' is a new section in BS 7671:2008.

1.7.1 Operation of protective devices in series
536.1 Regulation 536.1 sets out the requirements for coordination and states that where coordination of series protective devices is necessary to prevent danger and where required for proper functioning of the installation, consideration shall be given to selectivity and/or any necessary back-up protection.

Selectivity between protective devices depends on the coordination of the operating characteristics of two or more protective devices such that, on the incidence of fault currents within stated limits, only the device intended to operate within these limits does so. Regulation 536.1 also gives requirements for breaking capacity of a protective
536.4 device and Regulation 536.4 gives requirements for back-up protection.

By reason of the method of power distribution adopted, there may well be two or more overcurrent protective devices involved with a given fault current or overload; for example, where a main device protects a feeder to a distribution board with protection for a number of outgoing circuits, or where protection against overload and fault current is provided by two separate devices. There are then two aspects which need careful consideration: discrimination between operating characteristics, and coordination of their fault current breaking capacities and fault energy withstands.

1.7.2 Discrimination between devices
314.1 It is a matter of convenience and fitness for purpose of an installation that disconnection
536.1 interrupts the faulty or overloaded circuit only. It can also be a matter of safety.

Correct selection and comparison of the characteristics of each device will ensure that only the device electrically nearest to the cause of the overcurrent operates. The breaking capacity of the downstream device must therefore be suitable for the highest prospective current at its point in the circuit.

The operating characteristic of a given device lies within a band, the lower boundary of which constitutes a specified non-fusing or non-tripping characteristic, the upper boundary a specified fusing or tripping characteristic. The manufacturer can provide guidance or information such that a desk comparison between the non-operating limit of the upstream device and the operating limit of the downstream device will normally be sufficient to select devices to give correct discrimination.

When the prospective current is high and interruption times are of the order of 1 or 2 half-cycles it is unwise to attempt to assess discrimination by overlaying the characteristics. Devices when combined can behave very differently from when they are separate, and advice should be sought from the manufacturers, who can recommend suitable combinations. Publication PD IEC/TR 61912-2 *Low-voltage switchgear and controlgear – Overcurrent protective devices – Selectivity under overcurrent conditions* provides a comprehensive explanation of this subject.

536.2

As a rule of thumb, satisfactory discrimination between two similar types of fuse link is usually achieved when the downstream fuse has a rating not exceeding half that of the upstream fuse. For other devices a much higher ratio may be required, and in some circumstances satisfactory discrimination may not be possible.

Where discrimination is achieved, each protective device must comply with the requirements of Regulations 432.1 and 434.5.1 with regard to making and breaking prospective fault current at its point in the circuit.

432.1
434.5.1

1.7.3 Coordination of devices

It is sometimes not feasible or economical to provide a downstream device which has a sufficiently high fault breaking capacity.

430.3

A common example of this is the combination of a fuse or circuit-breaker with a motor starter. In general, motor starters are not designed, nor are they intended, to interrupt short-circuits or earth fault currents of a similar magnitude. However, in combination with the fault protection, they must be able to close on to the prospective fault current and to interrupt currents equal to the maximum overload capability of the associated motor. Fault protection for the starter circuit has to be provided by the fuses or circuit-breaker at the origin of the circuit, in compliance with Regulations 434.2 and 434.5.1.

435.2

434.2
434.5.1

The breaking capacity of the fault current protective device must be adequate to interrupt prospective fault currents, but its characteristics must also be such that it does not interrupt starting currents. Such characteristics are likely to be too high to operate on overload, either for the motor or to comply with Regulation 433.1.1 for the circuit conductors.

433.1.1

In motor circuit applications where circuit-breakers or gG type fuse links are used, the need to withstand motor starting currents usually dictates a higher current rating than would be selected on the basis of motor full-load current.

However, to meet this special role, circuit-breakers providing fault protection only may be used and extended ratings of motor circuit protection fuse links with gM characteristics are available, giving economies in both fusegear size and cost. These gM motor circuit fuse links have a dual rating: a maximum continuous rating based on the equipment in which they are fitted, for example a fuse carrier and base, and a rating related to the operational characteristics of the load.

In the type designation these two ratings are separated by the letter M. For example, 32M50 represents a maximum continuous rating of 32 A (governed by the associated fitment) and an operational characteristic of a 50 A fuse link.

430.3 Regulation 430.3 requires that the energy let-through of the fuse or circuit-breaker shall not exceed the withstand capability of the motor starter. In general, it is important that the overload relay does not initiate the opening of the starter contacts before the fault protection has had time to operate.

Where the fault protection is incorporated in one piece of equipment with the starter, it is the responsibility of the manufacturer to ensure that correct coordination is achieved.

435.2 Where the devices are separate, the manufacturer of the starter can provide guidance on the selection of a suitable fuse or circuit-breaker for fault current protection.

Motor starters and contactors are not generally self-protecting against the effects of short-circuit and therefore need to be associated with a short-circuit protective device (SCPD). In this particular case, test procedures according to BS EN 60947-4-1 recognise the difficulty of protecting sensitive devices from damage under short-circuit conditions. Thus a special case of conditional rating is obtained which allows two types of coordination with an SCPD.

▶ Type '1' coordination requires that, under short-circuit conditions, the contactor or starter shall cause no damage to persons or installations and may not be suitable for further service without repair or replacement of parts.
▶ Type '2' coordination requires that, under short-circuit conditions, the contactor or starter shall cause no damage to persons or installations and shall be suitable for further use. The risk of contact welding is recognised, in which case the manufacturer shall indicate the measures to be taken as regards the maintenance of the equipment.

These ratings can only be obtained by type-testing and thus the data for the selection of the SCPD must be obtained from the manufacturer of the contactor/starter, taking into account the rated operational current, rated operational voltage and the corresponding utilisation category.

Examination and maintenance of both devices is necessary after a fault.

The rated conditional short-circuit current of contactors and starters backed up by SCPDs, combination starters and protective starters is verified by short-circuit tests at two levels of prospective current:

1 at the rated conditional short-circuit current; and
2 an additional test made at a current 'r' (see Table 12 of BS EN 60947-4-1). The test current 'r' is considered a critical current for a contactor and the test ensures the performance of the contactor at this level.

Note: Further information about coordination between fuses and contactors/motor starters is given in IEC/TR 61459 *Low-voltage fuses – Coordination between fuses and contactors/motor-starters – Application guide.*

The overload relay in the starter is arranged to operate for values of current from just above full-load to the overload limit of the motor, but it has a time delay such that it does not respond to either starting currents or fault currents. This delay provides discrimination with the characteristics of the associated fuse or circuit-breaker at the origin of the circuit.

The starter overload relay can provide overload protection for the circuit in compliance with Regulation 433.1.1 on the basis that:

433.1.1

i its nameplate full-load current rating or setting is taken as I_n, and
ii the motor full-load current is taken as I_b, and
iii the ultimate tripping current of the overload relay is taken as I_2.

Where the overload relay has a range of settings then items **i** and **iii** should be based on the highest current setting, unless the settings cannot be changed without the use of a tool.

It is unlikely that the overload withstand of the circuit conductors will coincide with that of the motor, so overload settings should be determined by the lower requirement. Alternatively, an additional trip system, based on detection of motor temperature, may be used.

As a starter provides overload protection only, Regulation 433.2.2 permits it to be located anywhere along the run of the circuit from the distribution board, providing there is no other branch circuit or outlet between the starter and the motor.

433.2.2

1.7.4 Back-up protection

Another area where coordination is needed occurs when it is necessary to provide back-up for a fault current protective device with another one upstream of higher rating or higher fault current breaking capacity. It is not always satisfactory to evaluate the performance of the two devices by a comparison of their characteristics. For example, with short operating times such an approach would not take account of the effect of either device on the magnitude of the fault current, and their performance in series may be different from their individual performances.

434.5.1
536.4

Information should be obtained from the manufacturer of the downstream device as to the type of upstream fuse or circuit-breaker best suited to the prospective fault current.

Publication PD IEC/TR 61912-1 *Low-voltage switchgear and controlgear – Overcurrent protective devices* gives a comprehensive explanation of this subject.

Protection against overload

2

2.1 Introduction

Section 433

An *overload current* is a current the value of which exceeds the rated current of an electrically sound circuit. It may be caused by a user either deliberately or accidentally using more power than the circuit is intended for, or by a fault in equipment supplied by the circuit, or it may be caused by a characteristic of the load. It is not due to an electrical fault in the circuit itself.

Part 2

Damage by overload can take the form of accelerated deterioration of insulation, contacts and connections and deformation of some materials, thereby reducing the safe life and capability of equipment.

An overload caused by plugging in too many appliances may not be large, but it could last long enough to cause prolonged excessive heating of the conductors and probably of some of the accessories.

On the other hand, an overload caused by starting a motor may be several times the circuit rating, but its duration is not long enough to cause unacceptable overheating. However, where frequent motor starting or reversing at short intervals is expected, heating will be cumulative and the circuit components should be suitable for the higher duty.

552.1.1

The important characteristics of an overload are therefore the magnitude of the current and its duration.

2.2 Load assessment

Proper assessment of the load current I_b is essential for the safe application of these regulations.

433.1

This is reflected in the fundamental principles (Chapter 13) and includes consideration of the maximum demand on each distribution and final circuit.

133.2.2

Although overload protection is based on conductor temperature, it is equally important that all joints and connections are also capable of withstanding the currents and consequent temperatures expected.

132.6
132.7

Where the circuit supplies a single unvarying load or a number of such loads, I_b is obtained by simply summing all the loads concerned. In case of doubt, or for simple installations, this is the safest course.

Note: Refer to Appendix 4 of BS 7671 for method of installation and thermal insulation requirements. Also see section 3.7 of this Guidance Note regarding harmonics.

2.2.1 Diversity

311.1 In some cases it may be appropriate to recognise that not all the loads are operating at the same time and to apply a diversity factor. Determination of an appropriate diversity factor requires care and experience and account should be taken as far as possible of future expansion of the load or of change of duty.

GN1 Further advice on the determination of diversity factors is given in Guidance Note 1: *Selection & Erection*, Appendix H.

2.2.2 Varying loads

533.2.1 Where a load changes in a known manner it is possible to equate the design current I_b to a thermally equivalent constant load. Varying loads include both irregular pulses of load and regular or cyclic variations.

2.2.3 Thermally equivalent current

Where a circuit supplies a number of independent loads, or a fluctuating load, a simple procedure is to assess the value of the highest total load on the circuit and to substitute 433.1.1 this for I_b in Regulation 433.1.1.

This approach is adequate for circuits supplying small loads, where further design effort is not justified.

533.2.1 However, Regulation 533.2.1 indicates that regular repetitive loads (cyclic loads) should be assessed on the basis of a thermally equivalent constant load.

The benefits of such an approach can be extended to any varying load.

Descriptions of methods to calculate thermally equivalent loads are outside the scope of this Guidance Note. More information is given in the *Commentary on IEE Wiring Regulations*.

2.2.4 Small overloads of long duration

433.1 If I_b is equal to the sum of all expected loads, or to a thermally equivalent load, 523.1 there should be no overload which will result in the conductor temperature exceeding Table 52.1 its permitted working value. Overload protection then fulfils its proper function of protecting the circuit against events over which the designer has no control.

However, examination of the characteristics of many overload protective devices will reveal that, for practical reasons, they may not provide overload protection against currents just exceeding the rating of the associated conductors and equipment. For this reason it is important that every care is taken to avoid even small overloads of a protracted nature. (Small overloads are further discussed in section 2.3.6.)

Ageing and deterioration of insulation and connections increase rapidly at temperatures exceeding the rated values. The limits permitted assume that excess temperatures will be very infrequent and generally due to unforeseen situations. It is not practicable to place a general limit on the permissible number of overload events.

To avoid accelerated deterioration, it is essential that the design load of a circuit is high enough to include all foreseeable peak loads of a protracted nature.

It is never acceptable to use a fuse or a circuit-breaker as a load limiting device. This point is important because a small overload current may not be sufficient to operate the protective device, but could raise the temperature of the conductor above the rated value and prematurely age the cable.

The advice of the British Cables Association on cable life is: 'Estimating the life of a cable can only be approximate.'

There is no definitive or simple calculation method that can be used to determine the life expectancy of a fixed wiring cable.

Many factors determine the life of a cable, for example:

▶ mechanical damage
▶ presence of water
▶ chemical contamination
▶ solar or infrared radiation
▶ number of overloads
▶ number of short-circuits
▶ effects of harmonics
▶ temperature at terminations
▶ temperature of the cable.

Provided an applicable British Standard cable or equivalent has been selected taking into account operating temperature, installation conditions, equipment it is being attached to, and in accordance with appropriate regulations, it should meet or exceed its design life.

The design life of good quality fixed wiring cables is in excess of 20 years, when appropriately selected and installed. This design life has been assessed on a maximum loading – that is, the cable running at the maximum conductor temperature for 24 hours a day and 365 days a year.

If an installation is not fully loaded all the time, the expected life of the cable would be greater than the design life of the cable.

There are many instances of good quality fixed wiring cables operating for in excess of 40 years; this is mainly due to the cables not being abused and being lightly or only periodically loaded.

2.3 Selection of protective device and conductor size

There are two regulations in particular which need to be considered when selecting an overcurrent device.

The first specifies certain relationships between: 433.1.1

I_b the load (design) current,
I_n the nominal current or current setting of the protective device,
I_z the current-carrying capacity (rating) of the circuit conductors, and
I_2 the current causing effective operation of the device.

The relationships between I_b, I_n, I_z and I_2 can be set out as equations, or in diagrammatic form as in Figure 2.1.

i $I_b \leq I_n$
ii $I_n \leq I_z$
iii $I_2 \leq 1.45 \times I_z$.

Note that the load current, I_b, is the starting point of Regulation 433.1.1.

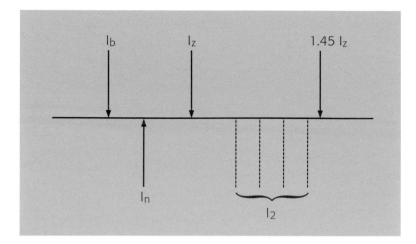

▼ **Figure 2.1**
Overload protection

Conditions **i** and **ii** provide the basic relationships for the selection of a suitable size or setting of the overload device ($I_b \leq I_n$), and an adequate conductor ($I_n \leq I_z$). The significance of selecting the conductor rating in this way is that it provides a link between the operating characteristic of the device and the conductor rating and paves the way for condition **iii**.

The alternative term 'current setting' for I_n applies to devices which have adjustable characteristics, when the nominal size might be larger than the current-carrying capacity of the conductor.

533.1.1 Where the device is adjustable, the means of adjustment should be sealed, or constructed so that alteration cannot be made without the use of a tool. A non-adjustable device should be used where untrained personnel may have access.

433.1.1 Condition **iii** provides the overload protection required by Regulation 433.1.1, but note that it is concerned primarily with the conductors. Protection for electrical equipment in the circuit, including the overload device itself, should be covered by careful selection of equipment, made to a standard quoted in the Regulations or to an equivalent.

The factor 1.45 in condition **iii** is based on a combination of experience and investigation. This has shown that the types of cable considered in BS 7671 can safely withstand a small but undefined number of periods at excess temperatures corresponding to currents not greater than 1.45 times their current-carrying capacity. However, such currents must not persist for long periods.

433.1.2 The second regulation always to be considered with Regulation 433.1.1 is Regulation 433.1.2. This states that, for the devices specified in Regulation 433.1.2 (these are the devices commonly used in the UK with the particular exception of rewireable fuses), the requirement for $I_2 \leq 1.45\, I_z$ will be met if $I_n \leq I_z$.

2.3.1 Selection of conductor size, and value of I_z

433.1.1
Appx 4
Values for I_z, the current-carrying capacity of conductors, can be obtained from Appendix 4 of BS 7671. Attention is drawn to paragraphs 3, 4 and 5 of that appendix where the selection of a conductor size to satisfy Regulation 433.1.1 is described.

Only the magnitude of the load current is specified in Regulation 433.1.1; the duration is taken care of by using a protective device having a suitable time/current characteristic.

433.1.1

A device complying with one of the British Standards specified in Regulation 433.1.2 has a characteristic designed so as to break an overload current within the limit of 1.45 times the conductor rating as required by condition **iii**, provided that it has been chosen so as to comply with condition **ii**.

433.1.2

Furthermore, its characteristic is such that the duration of an overload will be limited to a safe value.

Where fuses are used it is important to note that only the gG type cartridge fuse to BS 88-2.1 will automatically provide overload protection complying with both conditions **ii** and **iii** of Regulation 433.1.1. A 'gG' type fuse refers to a current-limiting fuse link which is intended for general purpose use, particularly where overload protection is required. It should be distinguished from types 'gM' (for motor circuits) and 'a' (partial-range breaking capacity) fuse links, which are not intended for overload protection.

433.1.1

Regulation 433.1.3 applies when semi-enclosed (rewireable) fuses to BS 3036 are used because they operate when the overload current is twice the nominal current rating of their fuse element.

433.1.3

Correct protection can be obtained by modifying condition **ii** so that the nominal current of such a fuse does not exceed $1.45/2 = 0.725$ times the current rating I_z of the associated conductor. The practical effect of this modification is to increase the size of conductor required for a given load.

Further comments on the use of BS 3036 fuses are made in Appendix 4 of BS 7671.

Appx 4

Lastly, attention is drawn to buried cables, for which suitable ratings are now included in BS 7671:2008:

1 Rating factor C_C must always be applied where overload protection is required for a cable in a duct in the ground or buried direct, so as to comply with condition **iii** of Regulation 433.1.1. This relates to the temperature values associated with conductor overload conditions (C_C is 0.9, or where the cable is protected by a semi-enclosed fuse C_C is 0.653). See Appendix 4.

433.1.4

2 The soil thermal resistivity must be considered. Generally in the UK for mixed soil/rubble conditions a value of 2.5 K.m/W may be used with a rating factor of 1. See paragraph 2.2 of Appendix 4.

523.3

Appx 4

2.3.2 Avoidance of unintentional operation of circuit-breakers
The following typical manufacturer's guidance is reproduced with the kind permission of Hager, a member of BEAMA.

> The unintentional operation of circuit-breakers is most commonly known as nuisance tripping and care must be taken in the selection of circuit-breakers to prevent their unintentional operation.

533.2.1
433.1

Regulation 533.2.1 states that: 'The rated current (or current setting) of the protective device shall be chosen in accordance with Regulation 433.1. In certain cases, to avoid unintentional operation, the peak current values of the loads may have to be taken into consideration'.

The peak current values of a load are the starting characteristics, which may include inrush current. The load characteristics should be compared with the minimum tripping current of the circuit-breaker.

As illustrated in Figure 2.2, inrush peak load current should be compared with the minimum peak tripping current of the circuit-breaker to ensure that unwanted tripping is avoided. The inrush peak current value is obtained from the manufacturer of the load equipment, and the minimum peak tripping current calculated by multiplying I_n by the lower multiple of the instantaneous trip and then by 1.414 (crest factor).

Different types of circuit-breaker (cb) to BS EN 60898 (which has replaced BS 3871) are available which operate at various values of short duration overcurrent (instantaneous trip current, this being a multiple of I_n) as shown in Table 2.1. See also Table 2.2, which gives circuit-breaker minimum peak tripping currents.

▼ **Figure 2.2**
Circuit-breaker and load current characteristics

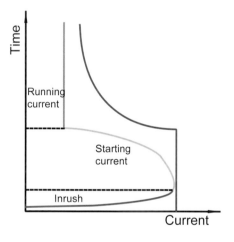

The different types of circuit-breaker available can be split into two categories.

Circuit-breakers for household and similar applications to BS EN 60898 (formerly known as mcbs) and RCBOs to BS EN 61009-1

There is a wide range of circuit-breaker characteristics that have been classified according to their instantaneous trip performance, and Table 2.1 gives some information on the application of the various types available. These limits are the maximum allowed for miniature circuit-breakers to BS 3871 (now withdrawn) and circuit-breakers to BS EN 60898-1, but it should be noted that manufacturers may provide closer limits. The same types B, C and D also apply to RCBOs to BS EN 61009-1.

Note: An RCBO is a residual current operated circuit-breaker with integral overcurrent protection.

Low voltage switchgear and controlgear Part 2 circuit-breakers to BS EN 60947-2

Unlike circuit-breakers to BS EN 60898-1, circuit-breakers complying with BS EN 60947-2 do not have defined characteristics and manufacturers' data must be used. BS EN 60947-2:2006 includes Annex L, which is for circuit-breakers not fulfilling the requirements for overcurrent protection (cbis) derived from the equivalent circuit-breaker. A class X cbi is fitted with integral short-circuit protection, which may, on the basis of the manufacturer's data, be used in conjunction with a motor starter overload relay for short-circuit protection.

Type	Instantaneous trip current	Application
1*	2.7 to 4 I_n	Domestic and commercial installations having little or no switching surge.
B	3 to 5 I_n	
2*	4 to 7 I_n	General use in commercial/industrial installations where the use of fluorescent lighting, small motors etc. can produce switching surges that would operate a Type 1 or B cb. Type C or 3 may be necessary in highly inductive circuits such as banks of fluorescent lighting.
C	5 to 10 I_n	
3*	7 to 10 I_n	
4*	10 to 50 I_n	Not suitable for general use. Suitable for transformers, X-ray machines, industrial welding equipment etc., where high inrush currents may occur.
D	10 to 20 I_n	

▼ **Table 2.1**
Selection of cb or RCBO type

* No longer available but may be found in existing installations.

For guidance on the selection of the type and number of loads that can be switched simultaneously, reference to manufacturers' data is recommended. The following typical manufacturer's guidance is reproduced with the kind permission of Hager, a member of BEAMA.

Circuit-breaker type	Circuit-breaker rated current								
	6 A	10 A	16 A	20 A	25 A	32 A	40 A	50 A	63 A
B	26	43	68	85	106	136	170	212	268
C	43	71	113	142	177	223	283	354	446
D	85	142	226	283	354	453	566	707	891

▼ **Table 2.2**
Circuit-breaker minimum peak tripping currents (A)

Circuit-breaker frame	I_n	Tripping current (A)	
		min	max
H125D	–	–	–
H125H	16–125	–	1131
H250N	160	905	1810
	200	1131	2262
	250	1414	2828
H400N	320	1810	3620
	400	2262	4525
H630N	500	2828	5656
	630	3563	7126
H800N	800	4525	9050

▼ **Table 2.3**
Circuit-breaker tripping currents

For other settings use I_n x magnetic setting − 20% x 1.414.

Lighting circuit applications

For the protection of lighting circuits the designer must select the circuit-breaker with the lowest instantaneous trip current compatible with the inrush currents likely to develop in the circuit.

High-frequency (HF) ballasts are often singled out for their high inrush currents but they do not differ widely from conventional 50 Hz versions. The highest value is reached when the ballast is switched on at the moment the mains sine wave passes through zero. The HF system is a 'rapid start' arrangement, in which all lamps start at the same time. Therefore the total inrush current of a lighting circuit incorporating HF ballasts exceeds the usual values of a conventional 50 Hz system. Where multiple ballasts are used in lighting schemes, the peak current increases proportionally. Mains circuit impedance will reduce the peak current but will not affect the pulse time.

The problem facing the installation designer in selecting the correct circuit-breaker is that the surge characteristics of high-frequency ballasts vary from manufacturer to manufacturer. Some may be as low as 12 A with a pulse time of 3 milliseconds and some as high as 35 A with a pulse time of 1 millisecond. Therefore it is important to obtain the expected inrush current of the equipment from the manufacturer in order to find out how many HF ballasts can safely be supplied from one circuit-breaker without the risk of nuisance tripping. This information can then be divided into the minimum peak tripping current of the circuit-breaker.

Example
How many HF ballasts, each having an expected inrush current of 20 A, can be supplied by a 16 A type C circuit-breaker?

From Table 2.2, for a 16 A type C device the minimum peak tripping current is 113 A.

Therefore $113/20 = 5.65$

Hence, 5 such ballasts can be supplied by a 16 A type C circuit-breaker.

2.3.3 Cables for star-delta starters

When a motor is controlled by a star-delta starter (Figure 2.3) the ends of the three motor windings are extended from the motor terminals to the starter. The connections of the windings in star for start and delta for run are made at the starter.

552.1 The steady running current in each of the six conductors between a star-delta starter and its motor is 58 per cent of that through the conductors supplying the starter. However, motor cables are usually run as a group having six conductors so that their current-carrying capacity is 80 per cent of a single circuit three-conductor capacity.

It follows that the single circuit rating of motor cables should be at least 72 per cent (0.58/0.80) of that of the associated supply cables. Examination of rating tables shows that the empirical rule by which motor cables are half the cross-sectional area of the supply cable does not always meet this requirement. Neither does it take into consideration possible differences in cable type or installation conditions.

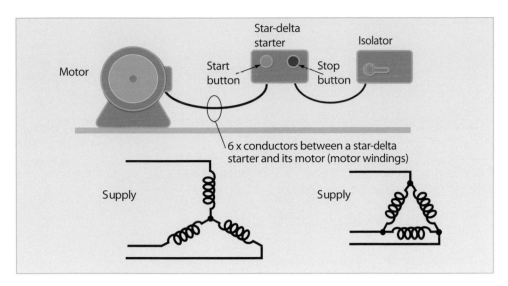

A simple procedure for the correct selection of motor cables, whereby the motor cables are chosen to have a rating, I_z, of at least 72 per cent of that of the supply cables, is illustrated by the following example, which uses the symbols and tables of Appendix 4 of BS 7671.

Appx 4

Example
Assume a full-load motor current of 37 A $= I_b$.

Supply cable to be three-core thermoplastic swa on a perforated tray. See Table 4D4A.

Table 4D4A

Ambient temperature 30 °C ($C_a = 1$).

See Appendix 4, Section 5.

Appx 4, Section 5

Motor cables to be single-core 70 °C thermoplastic in conduit. See Table 4D1A.

Table 4D1A

Ambient temperature 40 °C ($C_a = 0.87$). See Table 4B1.

Table 4B1

For the supply cable, Regulation 433.1.1 ($I_b \le I_n$) is met by a starter for which $I_n = 40$ A.

433.1.1

The cable size is then selected as in Appendix 4, 5.1.1.

Appx 4, 5.1.1

$$I_t \ge \frac{I_n}{C_a\,C_i\,C_c} = \frac{40}{1 \times 1 \times 1} = 40 \text{ A}$$

A three-core 6 mm^2 cable has a tabulated rating of $I_t = 45$ A (Table 4D4A, column 5).

Table 4D4A

If the delayed overload trips operate at, say, 50 A, condition (iii) of Regulation 433.1.1,

433.1.1

$$I_2 \le 1.45\,I_z \text{ where } I_z = C_a\,C_i\,C_c\,I_t$$

is met because $I_2 = 50$ A, and

$$1.45 \times 1 \times 1 \times 1 \times 45 = 65 \text{ A}.$$

For the motor cables, the procedure requires that

$$I_t \geq \frac{I_n}{C_a \, C_i C_c} \times 0.72 = \frac{40}{0.87 \times 1 \times 1} \times 0.72 = 33 \, A$$

Table 4D1A

$6 \, mm^2$ single-core cables in conduit, which have a single circuit tabulated rating, I_t, of 36 A, will be suitable (Table 4D1A, column 5).

If overload protection is satisfactory for the supply cables, the motor cables will also be protected. This can be demonstrated as follows:

$I_2 = 50$ (which is a line current)

For the motor cables, which are a double circuit in conduit,

$I_z = C_g \, C_a \, C_i \, C_c \, I_t$

$= 0.8 \times 0.87 \times 1 \times 1 \times 36 = 25 \, A$

$1.45 \, I_z = 1.45 \times 25 = 36.25 \, A$

The line equivalent of this current in the delta connected motor cables is:

$\sqrt{3} \times 36.25 = 63 \, A$

In this example, motor cables having a cross-sectional area (csa) of 50 per cent of the supply cable would be clearly inadequate. This is partly due to the use, in this example, of different types of cable and a higher ambient for the motor cables.

Even if the example had specified cables enclosed in conduit and the same ambient temperature for both runs of cable, the 50 per cent csa rule would fail, being saved only by the need to round up the size of the motor cables to the nearest larger standard size.

For higher loads and larger cables this rounding up is not always necessary and the 50 per cent rule is inadequate.

433.4

Appx 10

2.3.4 Cables in parallel

Appendix 10 of BS 7671

BS 7671:2008 includes a new, informative Appendix 10 from the IEC standard, which gives guidance on the protection of conductors in parallel against overcurrent. The appendix advises that overcurrent protection provided for conductors connected in parallel needs to provide adequate protection for all the parallel conductors.

433.4.1

For two conductors of the same cross-sectional area, conductor material, length and disposition arranged to carry substantially equal currents the requirements for overcurrent protection are straightforward.

433.4.2

For more complex conductor arrangements, detailed consideration should be given to unequal current sharing between conductors and multiple fault current paths.

Appendix 10 covers the following points concerning overload protection of conductors in parallel:

- The conditions where a single protective device can be used to protect all the conductors
- The current-carrying capacity (I_z) of the parallel conductors
- The current sharing between parallel cables
- Effects of resistive and reactive components
- Overload protection requirements where there is unequal current sharing between parallel conductors
- Formulae for calculating the design current for each conductor.

General

It is permissible to use only one device to provide overload protection for circuits composed of conductors in parallel, provided that the device will operate before damage occurs to any of the cables.

To achieve this it is necessary that the load current divides equally between the conductors.

As a first step this can be achieved by using conductors of the same length, cross-sectional area and material. It is then assumed that overload currents divide equally between the conductors and that I_z is the simple sum of the current-carrying capacities of the conductors. There are some further requirements, however, dependent on the type of cables involved, as the arrangement of the cables may introduce mutual heating between them and reduce their current-carrying capacities.

433.4

433.4.1
523.8

Single-core cables

For single-core cables, only certain cable arrangements provide reasonable current sharing. Examples are given in Figure 2.4. Other arrangements may provide acceptable current sharing, but this should be checked.

Multicore cables

For multicore cables, the conductor arrangements can be of two types:

a All the lines and neutral are carried in each cable; a circuit protective conductor is either included in each cable (in armoured cables it would be the armour) or, if the cables are not armoured, it may be run as a separate conductor. Satisfactory current sharing is easily achieved with this arrangement provided that cables are of the same size, type and length. The total current-carrying capacity is proportional to the number of cables, provided that any reduction in rating due to grouping is taken into account. (Note: even though the cables in parallel are in effect one circuit, the cables still need to be treated as though each was a separate circuit from a grouping point of view.) Refer to Appendix 4, Table 4C1 in BS 7671:2008.

Table 4C1

b For unarmoured cables only, it is feasible to use them as single-core cables by connecting all live conductors together. However, any circuit protective conductor should be a separate conductor and not a conductor in one or more of the cables. In the event of a cable fault, the fault current withstand of such a conductor would be inadequate. Further, currents induced in loops formed by such conductors will reduce the current-carrying capacity of the cables.

It is important to bear in mind that the total current-carrying capacity of cores paralleled in this way is not proportional to the number of cores. Unless better information is available it is prudent to expect four-core cables to carry only three times the rating of one core.

▼ **Figure 2.4**
Cable configurations for single-core cables in parallel to limit inductive effect on current sharing

2 cables per phase (equal current sharing)

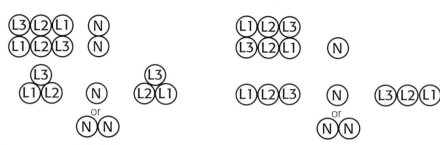

 = 1 neutral, = 2 neutrals, phase cables equally spaced or touching.
Trefoil groups touching.

3 cables per phase (no complete sharing possible with any configuration; example gives about 5 per cent unbalance, and up to 10 per cent circulating current on neutrals)

(*) = clearance D_e
(where D_e = the outside diameter of the cable)

(*) = clearance $3 D_e$
(where $3 D_e$ = 3 times the outside diameter of the cable)

1 neutral may be run in middle of each inter-group spacing.

4 cables per phase (flat formations provide equal sharing, no circulating current in neutrals, trefoils as for 3 cables)

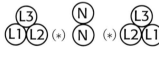

Cables touching or equally spaced

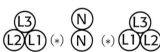 = 1 neutral, up to 4 positions may be used

(N) = 1 neutral, up to 4 positions may be used
If position is not used, space must be maintained

(*) = clearance D_e
(where D_e = the outside diameter of the cable)

Trefoils touching
Vertical clearance between trefoils $2 D_e$

2.3.5 Ring final circuits

433.1.5   Regulation 433.1.5 sets out the requirements for ring final circuits (Figure 2.5) with or
Appx 15   without unfused spurs. See also Appendix 15 of BS 7671. This regulation requires that the current-carrying capacity (I_z) of the cable is not less than 20 A, and that, under the intended conditions of use, the load current in any part of the ring is unlikely to exceed for long periods the current-carrying capacity of the cable. This is reinforced by
433.1   Regulation 433.1, which requires that every circuit be designed 'so that a small overload of long duration is unlikely to occur'.

This last point is important because a small overload current may not be sufficient to operate the protective device, but could raise the temperature of the conductors above the rated value and damage the cable. An explanation of what is a small overload and what is a long period is given in section 2.3.6.

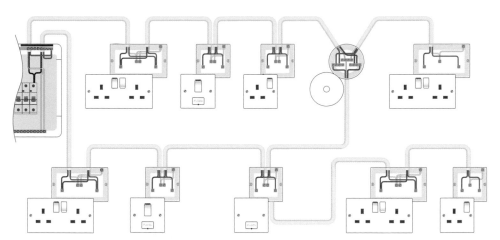

▼ **Figure 2.5**
Typical ring final circuit arrangement

Ring circuits are intended to provide large numbers of conveniently placed outlets, rather than a high load capability. The limitation on the floor area served by a ring circuit in household and similar installations (see the *On-Site Guide*, Appendix 8) is intended to provide a simple means of load limitation.

OSG, Appx 8

For household installations, a single 30 A or 32 A ring final circuit may serve a floor area of up to 100 m². However, careful consideration should be given to the loading in kitchens, which may require a separate circuit. See Figure 2.6 on kitchen appliance demands. Socket-outlets for washing machines, tumble dryers and dishwashers should be located so as to provide reasonable sharing of the load in each leg of the ring, or consideration should be given to a suitable radial circuit.

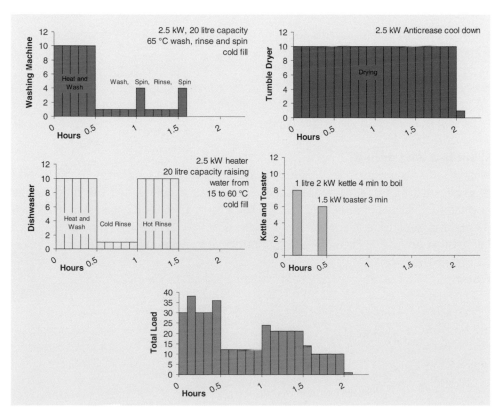

▼ **Figure 2.6**
Kitchen appliance demands (amperes)

Appx 15 Immersion heaters, or permanently connected heating appliances forming part of a comprehensive electric space heating installation, should be supplied by their own separate circuit.

Table 4D5A in BS 7671:2008 gives the current-carrying capacity of 70 °C thermoplastic (PVC) insulated and sheathed flat cable with protective conductor, which is often used in household installations.

For other premises or uses, a careful assessment should be made of the maximum load, its time diversity and its distribution around the ring, which may not be related to floor area. This is essential for non-domestic use because the circuit protection must not be used as a load limiter.

2.3.6 Load assessment
What is a small overload?
On the question of what is a small overload, guidance can be taken from the BS 88 fuse standard and the BS EN 60898 and BS EN 61009-1 circuit-breaker standards. For a BS 88 fuse, the 'non-fusing' current is 1.25 times the fuse rating, and for BS EN 60898 circuit-breakers and BS EN 61009-1 RCBOs the 'non-tripping' current is 1.13 times the rating of the circuit-breakers. (The non-fusing and non-tripping currents are the currents that the fuse or circuit-breaker must carry for the conventional time without operating.) The fusing current is 1.6 times the fuse rating, and the tripping current is 1.45 times the circuit-breaker rating. (These are the currents at which the fuse or circuit-breaker must operate within the conventional time, which is 1 hour for fuses and circuit-breakers rated up to and including 63 A.) For the tests, the fuse starts from cold, the circuit-breaker starts from a heated condition.

433.1.1 There is also a test under different conditions where the fuse must operate within the conventional time at 1.45 times its rating. BS 7671 accepts the 1.45 factor and hence accepts that an overload can be tolerated for the conventional time, that is, 1 hour for a 32 A fuse.

Circuit-breakers to the standards above must operate within the conventional time (i.e. within 1 hour) at 1.45 times their rating.

433.1 Every circuit must be designed so that a small overload of long duration is unlikely to occur. Even a small long-term overload of 6 per cent on a 90 °C rated cable can result in halving the expected life of that cable.

What is a long period?
A long period cannot be less than the 'conventional operating time' (1 hour) of a fuse because that is already allowed. It is also worth noting that the time/current **Appx 3** characteristic curves in Appendix 3 of BS 7671 for a 32 A BS 88 fuse stop at around 1 hour. It is reasonable to suggest that a 'long period' is a time exceeding that for which information on fuse or circuit-breaker performance is given, that is, exceeding 1 hour.

Note: BS 88 fuses are gradually being superseded by BS EN 60269 fuses.

2.3.7 Mineral insulated cables and ring final circuits

Regulation 433.1.5 gives a minimum cross-sectional area for both line and neutral conductors of 2.5 mm², except for two-core mineral insulated cables to BS EN 60702-1, for which the minimum is 1.5 mm². Concern has been expressed that the 73 °C operating temperature of mineral cables (for a 70 °C sheath temperature) might exceed that allowed for accessories by the appropriate standard (a temperature of 70 °C is assumed in the type tests of the standards). The current-carrying capacity of two-core 1.5 mm² light duty mineral insulated cable is 23 A (see column 2 of Table 4G1A of BS 7671). At this loading the estimated conductor temperature is 73 °C (sheath temperature 70 °C), which would overheat the accessory. This resulted in concerns regarding the advisability of applying the 20 A relaxation of Regulation 433.1.5.

However, assuming a maximum current of 20 A and that temperature rise is proportional to the square of the current, then temperature rise for a ring circuit with a maximum current of 20 A is given by:

$$\left(\frac{20}{30}\right) \times (73-30) = 32.51 \ °C$$

Assuming a maximum ambient temperature of 30 °C gives a conductor temperature of 62.51 °C, which is below 70 °C.

2.4 Omission of protection against overload current

Overload current protection may be omitted altogether under certain circumstances. However, it should not be omitted, or its sensitivity reduced, without very careful consideration; see comments relating to Section 433 in section 1.4 of this Guidance Note.

Regulation 433.3.3 allows the omission of devices for protection against overload for circuits supplying current-using equipment where unexpected disconnection of the circuit could cause danger or damage. The regulation gives the following six examples:

Note: In such situations consideration should be given to the provision of an overload alarm.

433.1.5

Table 4G1A

433.3.3

(i) The exciter circuit of a rotating machine.

(ii) The supply circuit of a lifting magnet.

Electromagnets are used, for example, in scrapyards to lift and carry loads. If such a magnet is de-energized while in operation this could cause damage or injury.

(iii) The secondary circuit of a current transformer.

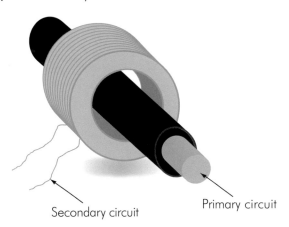

Secondary circuit   Primary circuit

The secondary circuit of a current transformer must be kept closed when current is flowing in the primary circuit. If not, the peak value of emf induced in the secondary circuit will be many times the nominal value, and could be high enough to cause breakdown of the insulation and danger.

(iv) A circuit supplying a fire extinguishing device. For example, a sprinkler system.

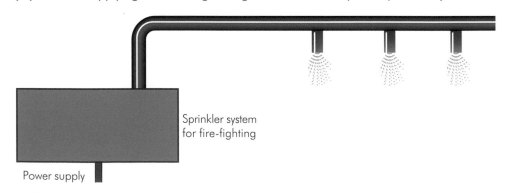

Sprinkler system
for fire-fighting

Power supply

(v) A circuit supplying a safety service, such as a fire alarm or gas alarm. For example, a supply to a fire alarm control panel.

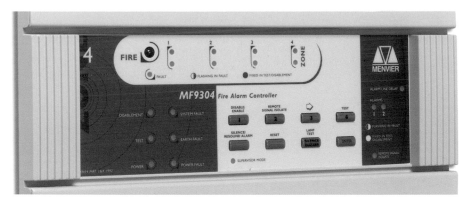

[Photograph courtesy of Menvier.]

(vi) A circuit supplying medical equipment used for life support in specific medical locations where an IT system is incorporated. Examples of life-support equipment include infusion pumps, ventilators and dialysis machines.

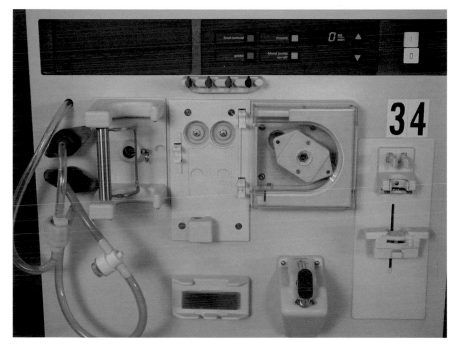

[Photograph courtesy of www.pemed.com]

433.3.1 Regulation 433.3.1 gives a relaxation where devices for protection against overload need not be provided. Such circumstances are:

(a) For a conductor situated on the load side of the point where a reduction occurs in the value of current-carrying capacity, where the conductor is effectively protected against overload by a protective device placed on the supply side of that point.

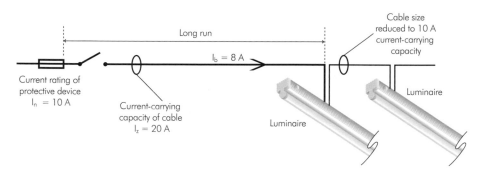

The diagram above shows a circuit in which a larger cable than necessary (from a current-carrying point of view) is used to reduce voltage drop, where the route to the first luminaire from the distribution board is a long run. The cable from the first luminaire supplying the remainder of the luminaires on the circuit is then reduced, but is protected against overload by the 10 A device.

(b) For a conductor which, because of the characteristics of the load or supply, is not likely to carry overload current, provided that the conductor is protected against fault current according to Section 434. For example, a domestic shower or immersion heater.

Section 434

(c) At the origin of an installation where the distributor provides an overload device and agrees that it affords protection to the part of the installation between the origin and the main distribution point of the installation where further overload protection is provided. This would normally apply to meter tails, as shown in the diagram below.

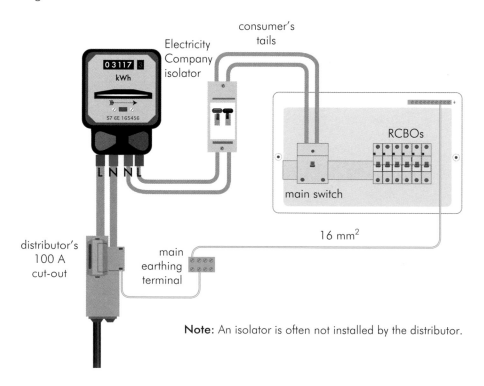

Note: An isolator is often not installed by the distributor.

Protection against fault current 3

3.1 Types of fault to be considered

Section 434

The Regulations require that each circuit should be provided with a device capable of breaking any fault current flowing in that circuit before it causes danger due to thermal or mechanical effects.

131.4
131.5
430.3
432.3

Fault currents result from an insulation failure or the bridging of insulation by a conducting item.

There are two types of fault:

▶ a short-circuit, where the current flows between live parts (which may include the neutral)
▶ an earth fault, where the current flows between a live part and earth. The latter includes any conductive part or conductor which is connected to the earth terminal or is substantially in contact with the mass of Earth.

The types of damage envisaged include:

▶ overheating and mechanical damage to conductors, connections and contacts
▶ deformation and deterioration of insulation leading to future electrical breakdown
▶ risk of ignition of materials adjacent to a conductor, due to its high temperature or to prolonged arcing at the fault or at a break in the circuit caused by the fault current.

3.2 Nature of damage and installation precautions

In general, fault currents are much larger than either load currents or overloads and the amount of damage which can be done is so great that rapid interruption is essential, as Figure 3.1 demonstrates.

BS 7671 requires that fault energy be limited to a value which will not cause a temperature rise sufficient to damage insulation (see Regulations 434.5.2 and 543.1.3), and appropriate temperature limits are given for live conductors in Table 43.1 and for protective conductors in Tables 54.2 to 54.6. Table 54.6, for bare conductors, is based on general considerations with regard to damage to the conductor or to unspecified materials which may be in contact or nearby. Damage means excessive deformation or reduction in thickness of the insulation.

434.5.2, 543.1.3
Table 43.1
Tables 54.2 to 54.6

The temperature limits given in these tables are based on the physical properties of the insulating materials, together with results of tests made on cables supported so that mechanical forces on the insulation would be representative of the values likely to occur in practice.

▼ **Figure 3.1**
Panel under test conditions showing the consequences of low to medium short-circuit faults [reproduced with the kind permission of Schneider Electric]

3.2.1 Installation precautions

134.1.1

Chap 52

Poor installation methods or adverse installation conditions may result in the imposition of greater forces, with the risk that damage may be excessive. For this reason, good workmanship is essential when installing cables, and methods of support should comply with those given in Chapter 52 of BS 7671.

522.8.3

In particular, careful attention should be paid to the minimum installation radii recommended in appropriate cable standards; the avoidance of excessive force during installation; contact with sharp edges; over-tight supporting clamps; unsupported lengths; leaving a cable in tension; and all situations where local stressing might occur.

3.2.2 Expansion forces

522.8.4
522.8.5

All conductors expand lengthways when heated, and cable support should be such that this expansion can be accommodated as uniformly as possible along the route. Supports and cable ties should be uniformly spaced and, in the case of larger cables, these should be set in a slightly sinuous path so that a small lateral movement is possible. Otherwise, expansion may accumulate at one point, causing severe bending, or at the ends, where it may overstress the terminations.

Larger sizes of conductor, roughly 50 mm^2 or more, are capable of exerting considerable longitudinal expansion forces and, unless the expansion is accommodated uniformly along the route, these forces may cause serious distortion of the conductor tails at terminations or mechanical overstressing of the connections.

Local restriction of larger cables at bends, particularly at the smaller radius bends, will produce high local stresses on both insulation and sheath, with an increased risk of damage.

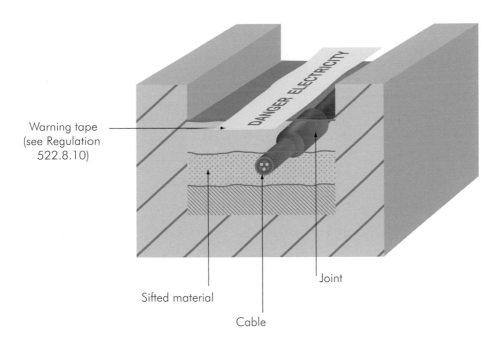

▼ **Figure 3.2**
Jointed cable buried
directly in the ground

Warning tape
(see Regulation
522.8.10)

Sifted material

Cable

Joint

Cables buried directly in the ground (Figure 3.2) have no opportunity for expansion relief by lateral movement, and expansion forces are applied directly to joints and terminations. Cable and joint makers should be consulted to confirm that the installation can withstand the short-circuit temperatures envisaged.

Allowance must be made for the thermal expansion and contraction of long runs of steel or plastic conduit or trunking, and adequate cable slack provided to allow free movement. The expansion or contraction of plastic conduit or trunking is significantly greater than that of steel for the same temperature change.

For busbar trunking systems, BS EN 60439-2 identifies a busbar trunking unit for building movements. Such a unit allows for the movement of a building due to thermal expansion and contraction. Reference should be made to the manufacturer since requirements may differ with respect to design, current rating or orientation of the busbar trunking system (i.e. riser or horizontal distribution).

522.15.1

Cables crossing a building expansion joint should be installed with adequate slack to allow movement, and a gap left in any supporting tray or steelwork. A flexible joint should be provided in conduit or trunking systems.

3.2.3 Electromagnetic forces
Compliance with the temperature limits in Table 43.1 may not avoid the risk of damage due to electromagnetic forces.

Table 43.1

These forces are unlikely to be important for circuits where fault current protection complying with Regulation 435.1 and rated at not more than 100 A is provided, but can become increasingly significant as fault currents increase above about 15 kA rms or about 20 kA pk. It should be borne in mind that an initial peak current approaching 1.8 times the rms value is possible when a fault is electrically near to a large transformer or generator source (see Figure 3.3).

435.1

The cut-off characteristics of current-limiting fault protective devices will act to restrict such high initial peak currents and the forces they would produce. Information can be obtained from the manufacturer of the device.

▼ **Figure 3.3**   Sources of electricity supply [reproduced with the kind permission of John Lewis Partnership]

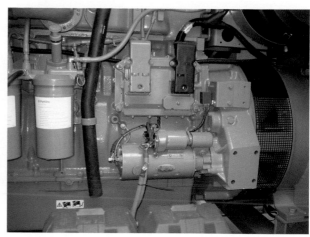

a Transformer source

b Generator source

It can be assumed that *unarmoured* multicore cables of all sizes are strong enough to withstand electromagnetic bursting forces (i.e. mechanical forces, forcing the conductors apart) where protection is coordinated with cable size in compliance with Regulation 434.5.2, when the fault current protection device is of the current-limiting type. Cable makers should be consulted where let-through currents are likely to be in the region of 25 kA rms or greater.

434.5.2

It can be safely assumed that all *armoured* multicore extruded insulation cables are strong enough to withstand bursting forces without damage, whatever type of protective device is employed. However, where currents may exceed 25 kA rms, it is wise to consult the cable maker.

521.5.1

In the case of single-core cables, electromagnetic forces act to move the cables apart. The effect is dependent on cable size, type and spacing (the closer the cables, the greater the forces), on the support provided by fixings and on the peak value of the current. Fault currents well in excess of 20 kA pk may result in severe bending of the cables, and with higher currents there is the possibility of broken fixings. To reduce the risk of damage, cables should be strongly bound together at frequent intervals to resist electromagnetic forces, but should be supported or fixed at greater intervals to allow lateral movement and uniform relief of longitudinal expansion by snaking. Again, the cable maker can provide information on the type of binding and supports to use.

A weak point as regards both longitudinal expansion forces and electromagnetic forces is unsupported lengths of core at joints and terminations. Electromagnetic forces are highest in the region of the crutch of multicore cable terminations, i.e. where the conductors are closest. Adequate support, which may be in the form of a suitable gland or binding, to reinforce the end of the cable envelope and to support the cores at the crutch is desirable. Lateral support for cores should be considered where tails are long enough to bend and impose strain on terminals.

Adequate support must be given to cables in busbar chambers to prevent abrasion of insulation by electromagnetic movement.

3.3 Fault impedance and breaking capacity of protective device

It is conventionally assumed that the impedance of the actual fault is zero. This makes calculation of the value of fault current straightforward, because only the impedances of the circuit conductors need be taken into account. Such an assumption neglects any arc voltage and provides the highest expected value of fault current at a particular point in a circuit, referred to as the prospective fault current, and hence the maximum current-breaking capacity required of the device to be selected to interrupt the current.

132.8
432.1
432.3
434.1

It is important for this selection to be correct, not only because the successful interruption of the current avoids unnecessary risk of damage, which in higher current circuits can be very serious, but also because, in the extreme, if the device fails to interrupt the current it is likely to be destroyed and the duty of interruption passes to the next device upstream. This next device is usually chosen to detect a much higher magnitude of fault current and may not operate, or may only operate after considerable damage and risk of danger to persons or property has occurred.

Table 3.1 gives the rated short-circuit breaking capacities for some widely used protective devices.

▼ **Table 3.1**
Rated short-circuit capacities

Device type	Device designation	Rated short-circuit capacity (kA)	
Semi-enclosed fuse to BS 3036 with category of duty	S1A	1	
	S2A	2	
	S4A	4	
Cartridge fuse to BS 1361 type I		16.5	
type II		33.0	
General purpose fuse to BS 88 Part 2.1		50 at 415 V	
Part 6		16.5 at 240 V	
		80.0 at 415 V	
Circuit-breakers to BS 3871	M1	1	
	M1.5	1.5	
	M3	3	
	M4.5	4.5	
	M6	6	
	M9	9	
Circuit-breakers to BS EN 60898* and RCBOs to BS EN 61009*		I_{cn}	I_{cs}
		1.5	(1.5)
		3.0	(3.0)
		4.5	(4.5)
		6	(6.0)
		10	(7.5)
		15	(7.5)
		20	(10.0)
		25	(12.5)

533.3

* Two rated short-circuit capacities are defined in BS EN 60898 and BS EN 61009:
   I_{cn} the rated short-circuit capacity (marked on the device).
   I_{cs} the service short-circuit capacity.
The difference between the two is the condition of the circuit-breaker after manufacturer's testing.
I_{cn} is the maximum fault current the breaker can interrupt safely, although the breaker may no longer be usable.

I_{cs} is the maximum fault current the breaker can interrupt safely without loss of performance. The I_{cn} value (in amperes) is normally marked on the device in a rectangle e.g. $\boxed{6000}$ and for the majority of applications the prospective fault current at the terminals of the circuit-breaker should not exceed this value.

For domestic installations the prospective fault current is unlikely to exceed 6 kA, up to which value I_{cn} will equal I_{cs}.

The short-circuit capacity of devices to BS EN 60947-2 is as specified by the manufacturer.

3.4 Position of fault current protection and assessment of prospective current

434.2 With certain exceptions, fault current protection is required at the origin of each circuit or, more generally, where there is a reduction in the size of conductor and hence in fault current withstand capacity.

3.4.1 Assessment of fault current

533.3 At the origin of a circuit, or point of reduction in conductor size, it is necessary to calculate both the highest and the lowest values of prospective fault current. The operating time and energy let-through (I^2t) of a protective device depends on the current and it is necessary to check that conductors being protected can withstand the higher of the two I^2t let-through values. It is a requirement that the device is capable of safely interrupting the highest prospective current and that it will operate correctly for the lowest prospective fault current.

The highest value of prospective fault current is calculated assuming a fault immediately on the load side of the device. The impedances concerned are those upstream of the device, including that of the supply to the installation, and should include, if significant, the impedance of the device itself.

The lowest value of prospective fault current is calculated on the assumption that a fault occurs at the load end of the circuit or at the input to the next downstream fault current protective device.

In a single-phase circuit the line to earth prospective fault current may be greater than that for a line to neutral fault.

3.4.2 Fault current and impedance at origin of installation

313.1 BS 7671 requires the prospective short-circuit current at the origin of the installation to be determined.

Engineering Recommendation P25/1 *The short-circuit characteristics of Public Electricity Suppliers low voltage distribution networks and the co-ordination of overcurrent protective devices on 230 V single-phase supplies up to 100 A*, which was published by the Electricity Association and is now available from the Energy Networks Association, indicates that at a service tee-off of a public supply network a typical maximum value of 16 kA may be assumed for single-phase supplies where the distributor has provided a cut-out rated up to 100 A. Engineering Recommendation P26 *The estimation of the maximum prospective short-circuit current for three-phase 415 V supplies*, which is also available from the Energy Networks Association, is applicable to three-phase supplies and quotes somewhat higher values.

From information on the length of service cable on private property, estimates can be made of the attenuation (reduction) in fault level from the figure provided by the distribution company. Chapter 6 provides advice on this. However, for domestic premises consumer units and fuseboards with a conditional fault rating of 16 kA can be used, so further calculation is not necessary. Consumer units and fuseboards to BS EN 60439-3 Annex ZA (Specification for particular requirements for consumer units, etc. complete with protective devices) are able to withstand the fault current for prospective fault levels up to 16 kA when the electricity distributor's fuse is a type II fuse to BS 1361 rated at no more than 100 A. The standard 100 A fuse installed by electricity distribution companies will meet these requirements. Whilst the individual overcurrent devices in the consumer unit or distribution board may not interrupt the fault current at high fault current levels, they will be able to carry the currents until the distributor's fuse operates. Clearly, it is better if, as is usually the case, the fault rating of the fuse or circuit-breaker in the distribution board or consumer unit is of sufficient rating to clear the fault, particularly faults downstream of the consumer unit. Fault levels are rarely as high as 16 kA and rapidly decrease within the installation, so in practice the fuse or circuit-breaker in the consumer unit will clear faults.

530.3.4

The Engineering Recommendations include information on the attenuation in fault current provided by the service cable to the consumer's installation, leading to much reduced values of short-circuit current at the service cut-out. The extent of the reduction depends on the size (csa) of cable and its length.

This reduction in short-circuit current by the service cable does not apply to installations in high load density areas of major city centres.

For other situations the distributor should be consulted or, if the source is a consumer's transformer, the manufacturer's impedance data should be applied.

The calculation of fault current at the origin of the installation, and hence of the impedance of the source, should take into account contributions, if any, from sources additional to the public or external supply. Such additional sources may take the form of local generation or of large motor loads connected electrically close to the supply point.

551.7.1

The lowest prospective value relates to a fault at the downstream end of the conductors protected by the device, notionally at the input to the next distribution board or at the terminals of the current-using equipment fed by that circuit.

As stressed earlier (section 1.1.3), protection of conductors in accordance with Chapter 43 may not provide protection for equipment connected to those conductors. Where a final circuit supplies a known single load it may be feasible to select a device whose characteristics will also protect the equipment, but reference to the equipment manufacturer may be necessary. Alternatively, equipment protection should be included at or in the equipment itself.

This aspect is of importance where non-fused plugs are used, such as industrial type to BS EN 60309-2. If the designer intends to incorporate such plugs and sockets he or she would require to establish the sizes and types of flexible cords likely to be used. There may be a case for limiting the current rating of the circuit, or setting a minimum csa or maximum length of cord to be used.

3.4.3 Relocation of fault current protection

434.2 Fault current protection may be located on the load side of the position required by
434.2.1 Regulation 434.2 provided that the conditions given in Regulation 434.2.1 are met.

The three conditions in Regulation 434.2.1 are intended to minimise the risk of a fault occurring and, if a fault should occur, the risk of fire and danger to persons or property. The following measures are envisaged:

1   The use of conductor lengths that are as short as possible
2   The reduction as much as is reasonably practicable of the risk of insulation breakdown, by choice of route, provision of adequate support and enclosure or barriers, together with the addition of supplementary insulation separately over each conductor
3   Consideration of the effects of mechanical damage, including conductor displacement, either from external mechanical forces or from electromagnetic forces in the event of a fault downstream
4   Although the prime object should be to avoid faults greater than the withstand of the conductors, locating or enclosing the conductors so that, in the event of a fault, risk of fire or danger to persons or property by the emission of flame, arcs or hot particles is prevented so far as is reasonably practicable.

411.3.1.1 Where metallic enclosures or barriers are used they are, by definition, exposed-conductive-parts and must be connected to the circuit protective conductor.

It is not satisfactory to assume that small conductors are safely expendable. If such a conductor is destroyed, the arc products will contain copious quantities of ionised gases which may facilitate the persistence and flashback of an arc to larger conductors, where arcing is almost certain to do considerable damage. Where fusing of small conductors cannot be provided, the run of the conductors should be carefully designed so as to maximise the chances of containment in the event of a fault.

A small conductor, such as one supplying instrumentation from a large conductor, can be connected through a small fuse unit of adequate fault current breaking capacity mounted directly on the large conductor.

▼ **Figure 3.4**   Interconnections between switchgear items

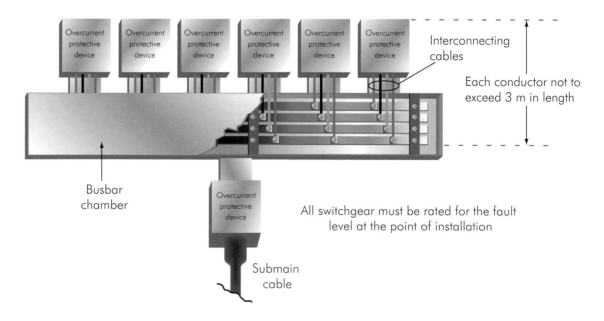

Regulation 434.2.1 is of special importance for interconnections between switchgear (Figure 3.4) and for temporary connections made to equipment where there may not be room for the installation of a suitable fault protective device at the tap-off point. In this latter situation the original integrity of the switchgear assembly and enclosure as regards protection against entry of foreign substances (Regulation 512.2), compliance with Chapter 41 'Protection against electric shock' and Chapter 42 'Protection against thermal effects' must be maintained.

434.2.1

The specification for type-tested and partially type-tested assemblies requires that the manufacturer shall specify in his documents or catalogues the conditions, if any, for the installation, operation and maintenance of the assembly and the equipment contained therein.

If necessary, the instructions for the transport, installation and operation of the assembly must indicate the measures that are of particular importance for the proper and correct installation, commissioning and operation of the assembly.

In addition, where necessary, the above mentioned documents must indicate the recommended extent and frequency of maintenance.

If the circuitry is not obvious from the physical arrangement of the apparatus installed, suitable information must be supplied, for example wiring diagrams or tables.

Regulation 434.2.2 applies where fault protection upstream of the point where a conductor cross-sectional area is reduced also provides fault protection complying with Regulation 434.5.2 for the smaller conductor. Further fault protection may then be provided at any point downstream of the conductor change (cascading), for example to provide discrimination but not primarily to protect the conductor.

434.2.2

434.5.2

3.5 Omission of fault current protection

Fault current protection may be omitted altogether under certain circumstances. However, fault current protection should not be omitted, or its sensitivity reduced, without very careful consideration; see comments relating to Section 434 in section 1.4 of this Guidance Note.

Regulation 434.3 allows the omission of devices for protection against fault current in the following four circumstances, provided that the conductor that is not protected is installed in such a manner as to reduce to a minimum the risks of fault and of fire or danger to persons.

434.3

(i) For a conductor connecting a generator, transformer, rectifier or battery to its associated control panel. [Photographs reproduced with the kind permission of John Lewis Partnership.]

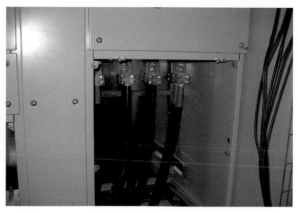

Transformer source

LV switchboard showing incoming supply cables from transformer

433.3.3 **(ii)** In a circuit where disconnection could cause danger for the operation of the installation, such as the secondary circuit of a current transformer.

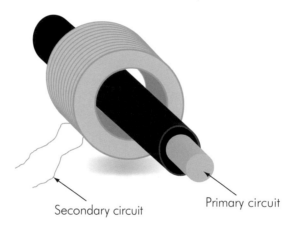

Secondary circuit

Primary circuit

The secondary circuit of a current transformer must be kept closed when current is flowing in the primary circuit. If not, the peak value of emf induced in the secondary circuit will be many times the nominal value, and could be high enough to cause breakdown of the insulation and danger.

Another example is the supply circuit of a lifting magnet, where unexpected opening of a circuit could cause greater danger than the fault current condition.

Electromagnets are used, for example, in scrapyards to lift and carry loads. If such a magnet is de-energized while in operation this could cause damage or injury.

(iii) Certain measuring circuits.

(iv) At the origin of an installation where the distributor installs one or more fault current devices and agrees that such a device affords protection to the part of the installation between the origin and the main distribution point of the installation where further protection against fault current is provided. An example would be fault protection of the meter tails in the diagram below by the distributor's cut-out fuse.

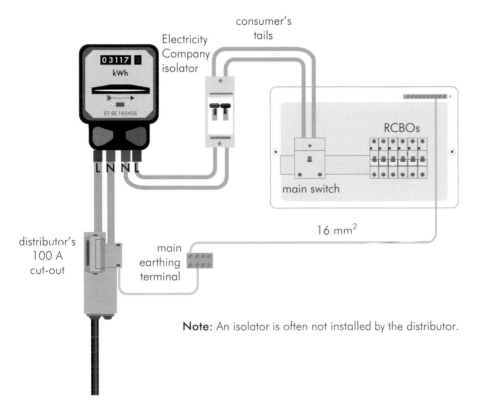

Note: An isolator is often not installed by the distributor.

Item (iv) of Regulation 434.3 must be considered together with item (iii) of Regulation 433.3.1 which also deals with the protection of cables between the outgoing meter terminals and the first point of overcurrent protection in an installation. In general, the distributor is able to make a satisfactory assessment of the maximum demand so that overload need not be a problem. The important issue will be fault current protection. Sizing and the method of installation of these cables must be agreed with the distributor. If Regulations 434.3(iv) and 433.3.1(iii) are not met then overcurrent protection, complying with Regulation 434.2.1, should be installed close to the origin.

434.3
433.3.1

434.2.1

3.6 The use of one device for both overload and fault current protection

Regulation 435.1 deals with a very common situation, where the same fuse or circuit-breaker provides both types of protection. The assumption relies on the inherent characteristics of the devices used, whereby correct overload protection, achieved by applying the conditions of Regulations 433.1.1 and 433.1.2, will involve devices having overcurrent characteristics which will ensure compliance with Regulation 435.1.

435.1

433.1.1
433.1.2

434.5.2 It may be deduced from the latter part of Regulation 435.1 that non-current-limiting devices may not provide the required characteristic. Doubt could arise when a device not described in Regulation 433.1.2 is used; a check must then be made, using the characteristic of the device chosen, for compliance with Regulation 434.5.2.

Fault current limitation refers to the characteristic of a device which is so constructed that interruption takes place in less than one half-cycle of current. Fuses to BS 88 provide this characteristic provided that they are suitably selected with regard to the value of prospective current.

Circuit-breakers can be of either type, the current-limiting type sometimes being referred to as 'fast acting'. A non-current-limiting type may be referred to as a 'zero point', 'current zero' or 'half-cycle' device.

Since there is no special marking to identify these types, it is necessary to refer to the manufacturer or trade literature.

3.7   Harmonics

523.6.1
523.6.3
431.2.1
431.2.3
533.2.2

Regulations 523.6.1 and 523.6.3 recognize the effect of triple harmonic currents in the neutral conductor and the need to take account of this. Regulations 431.2.1 and 431.2.3 are concerned with TN or TT systems. These regulations require overcurrent detection in the neutral conductor in a polyphase circuit, where the harmonic content of the phase currents is such that the current in the neutral conductor is reasonably expected to exceed that in the phase conductors. This detection shall cause disconnection of the phase conductors but not necessarily of the neutral conductor.

The reason behind these regulations is that certain equipment such as the switched mode power supplies of computers and discharge lighting produce third harmonics. This third harmonic content produces harmonic distortion. The problem occurs in three-phase circuits. For example, 50 Hz three-phase load currents, if balanced, cancel out in the neutral because of the 120-degree displacement of each phase (see Figure 3.5). Even if the load is not balanced the neutral conductor will only carry the out-of-balance current, which is less than the phase current. However, the third, and other triple harmonics combine in the neutral to give a neutral current that has a magnitude equal to the sum of the third harmonic content of each phase. The heating effect of this neutral current could raise the temperature of the cable above its rated value and damage the cable.

523.6.1
523.6.2
533.2.2

Therefore, consider a sub-main cable supplying a three-phase installation where there is likely to be a significant harmonic content; this is where computer loads or discharge lighting loads (or other loads which produce third harmonics) represent a significant proportion of the total load. Where the harmonic content of the phase currents is such that the current in the neutral conductor is reasonably expected to exceed that in the phase conductors, the regulations require overcurrent detection in the neutral and allowances must be made in the sizing of cables.

Appx 11 **Note:** BS 7671:2008 includes a new, informative Appendix 11 from the IEC standard, which gives guidance on the effect of harmonic currents on balanced three-phase systems. The appendix includes a table which provides derating factors for balanced three-phase loads and should be applied to four-core and five-core cables with four cores carrying current. The appendix also includes information on higher harmonic frequencies.

▼ **Figure 3.5** Phase displacement and third harmonics

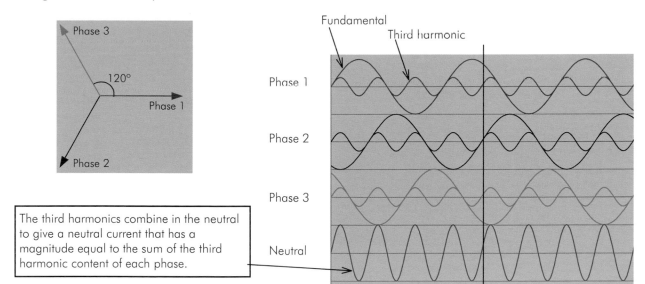

The third harmonics combine in the neutral to give a neutral current that has a magnitude equal to the sum of the third harmonic content of each phase.

Determination of fault current

4

Regulation 434.1 requires the prospective fault current to be determined at every relevant point of the installation either by calculation, measurement or enquiry.

434.1

Unless fault current is measured directly, its determination relies on measurement or calculation of relevant values of fault loop impedances.

4.1 Determination of fault current by enquiry

Engineering Recommendation P26 *The estimation of the maximum prospective short-circuit current for three-phase 415 V supplies*, which was published by the Electricity Association and is now available from the Energy Networks Association, deals with the estimation of the maximum prospective short-circuit current for three-phase 400 V supplies.

Engineering Recommendation P25/1 *The short-circuit characteristics of Public Electricity Suppliers low voltage distribution networks and the co-ordination of overcurrent protective devices on 230 V single-phase supplies up to 100 A*, which is also available from the Energy Networks Association, deals with the short-circuit characteristics of public electricity distributors' low voltage distribution networks and the coordination of overcurrent protective devices on 230 V single-phase supplies up to 100 A.

For 230 V single-phase supplies up to 100 A, the distributor will provide the consumer with an estimate of the maximum prospective short-circuit current at the distributor's cut-out, which will be based on Engineering Recommendation P25/1 and on the declared level of 16 kA at the point of connection of the service line to the LV distribution cable. The fault level will only be this high if the installation is close to the distribution transformer. However, because changes may be made to the distribution network by the distributor over the life of an installation, the designer of the consumer's installation must specify equipment suitable for the highest fault level likely.

Since the fault level of 16 kA is at the point of connection of the service line to the LV distribution cable, a reduction of this fault level is allowed for the service line on the customer's premises. This is because the service line on the customer's premises will not be changed without the customer's knowledge. There are some inner city locations, particularly in London, where the maximum prospective short-circuit current exceeds 16 kA, and the distributor would take account of this when advising the customer of the fault level.

For three-phase 400 V supplies, the distributor will provide the consumer with an estimate of the maximum prospective short-circuit current at the distributor's cut-out. This will be based on Engineering Recommendation P26, and will also be based either on the declared fault level of 18 kA at the point of connection of the service line to the

LV distribution cable or on the declared fault level of 25 kA at the point of connection to the distributor's substation.

Since the fault level 18 kA is at the point of connection of the service line to the LV distribution cable, again, a reduction of this fault level is allowed for the service line on the customer's premises.

In the event that the service cable is supplied from a distribution cable in the footpath on the far side of the road, the attenuation in fault level can only be applied from the footpath nearest to the property. This is because the distributor may, at some time in the future, install a distribution cable in the footpath nearest to the property from which to supply the service cable.

More information on the determination of fault current and detailed tables on the estimation of maximum prospective short-circuit current are given in the *Commentary on IEE Wiring Regulations*.

▼ **Figure 4.1** Typical values of fault level at distributor's cut-outs

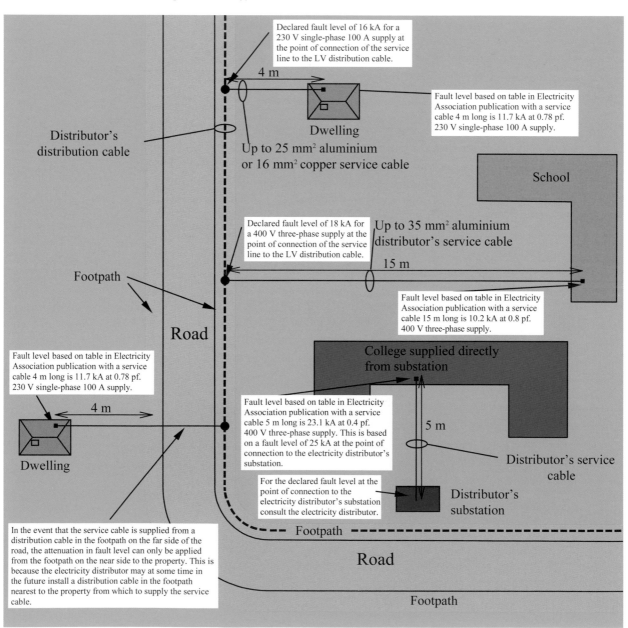

Figure 4.1 shows a number of consumers supplied via a distributor's cables, and the reduction in fault level at the distributor's cut-outs due to the length of service cable to each consumer's premises. Shown are typical values; refer to the distributor for actual value of fault level for a particular installation.

4.2 Measurement of fault current

Measurement of prospective fault current or circuit impedance requires special equipment. For circuits rated up to about 100 A, portable equipment is available and is generally used for the measurement of earth fault loop impedance. It operates by imposing a heavy load, usually greater than the rating of the circuit, for a very short duration, and measuring the consequent drop in voltage. From this data the upstream impedance and hence prospective fault current is derived.

For higher rated circuits the currents required to effect such a measurement accurately are proportionally higher and the equipment becomes cumbersome and expensive.

To achieve a satisfactory accuracy, test currents have to be high, and to avoid damage or unnecessary operation of protective devices the duration must be kept as short as possible, notionally only a few half-cycles.

The measurement involves some degree of danger, increasingly so at the higher currents. It should be undertaken only by skilled personnel with adequate equipment, operated according to the manufacturer's instructions, and taking appropriate safety measures.

For a larger installation an acceptable alternative is to disconnect the circuit from the supply and to measure its impedance with reduced voltage high-current test equipment. Where circuit impedance is very low this may prove to be the best, or only, way to achieve a satisfactory accuracy with safety.

With an external supply its impedance has to be ascertained from the distributor or assessed from published information and added to the values measured within the installation.

Measurement is a useful method when there is doubt as to the status of a calculated value, where the impedance of an existing circuit is required and the route is inaccessible or, for low current final circuits, as a convenient means of regular verification of earth loop impedance. Unless a complicated method of measurement is used, the resistive and reactive components of the impedance are not identified, so arithmetical addition of further impedances can only be approximate.

Accuracy of measurement, as opposed to accuracy of instrumentation, may not be as good as is sometimes thought, particularly for very extensive installations. Results may be affected by the presence of ferromagnetic materials, the effect of which can be sensitive to the value of the measuring current, near to or around conductors. This is particularly so for single-core cable circuits.

With earth fault loop measurements there is the possibility of ferromagnetic materials forming part of the loop, together with the uncertainty that subsequent operations by other trades or changes to a building structure may modify the effective path of fault currents.

Attention is drawn to the fallacy of reading the last one or two digits on a three- or four-digit test instrument display and omitting to overlay the result with the inherent instrument errors. The manufacturer's instructions should always be consulted before taking a reading, not only when a reading is felt to be unusual.

Where measurement indicates that the expected value of fault current is low for the type of supply concerned, consideration should be given to the possibility that the supply may be reinforced at some future date. Protective devices should be selected which will not become inadequate should reinforcement occur.

Measurement is usually done when the circuit is off-load, and the value obtained relates to conductors at ambient temperature.

4.3   Calculation of fault current

Derivation of circuit impedance and fault current values by calculation is the usual and sometimes the only way of providing design data. Calculation of fault current is possible only if all the impedances are known or can be estimated.

Provided that route lengths are accurate, better results are likely to be achieved with low current circuits where reactance can be ignored. Likely sources of error are connections and busbars for higher current installations, where an accurate calculation of reactance may be difficult or impracticable.

It should be borne in mind that where measurements have been made on installed circuits, values of prospective current have been reported to be lower than calculated values based on route lengths. The discrepancies have been attributed to unascertainable additional resistances and, particularly, reactances in equipment such as distribution boards and busbars. (See J. Rickwood, 'Fault Currents in Final Sub-Circuits', paper at IEE/ERA Conference on Distribution, Edinburgh, 20–22 October 1970.)

430.3
434.5.2

Errors where actual currents are lower than those estimated may be on the safe side as far as selection of breaking capacities and withstand of electromagnetic forces are concerned, but the repercussions for thermal damage limitation (see Regulations 430.3 and 434.5.2) are different. A lower fault current will correspond to a longer fault clearance time by the protective device and, in most cases, a greater energy let-through.

Fortunately, the situation is largely redressed by the fact that the actual operating currents of protective devices are generally somewhat lower than the maximum requirements of the relevant standards and of published data.

4.3.1   Highest value of fault current

The lowest impedance of a fault current path may be obtained by assuming that the fault occurs when the conductors are cold. In view of the fact that the highest value of fault current is usually required for the selection of a device with adequate breaking capacity, rather than to check on the fault current capacity of conductors, it is sufficient and on the safe side to neglect the increase in conductor resistance during a fault. Unless the conductors are located in a low ambient temperature, such as within a refrigerated area, resistance values tabulated for 20 °C may be used.

4.3.2 Lowest value of fault current

Contrary to expectation, the lowest value of fault current is used when considering thermal stresses on conductors as a result of a fault (Regulation 434.5.2). This comes about because the characteristics of most of the circuit protective devices considered in BS 7671 are such that the fault energy let-through (I^2t) is greater the lower the fault current. It follows that faults are assumed to occur when a circuit is at its maximum permissible working temperature and conductor resistances are high.

434.5.2

The lowest value of fault current is also used when confirming compliance with the maximum disconnection times required for shock protection, and the longest likely duration of the fault is to be checked against the requirements of Regulation 411.3.2.2, 411.3.2.3 or 411.3.2.4.

411.3.2.2
411.3.2.3
411.3.2.4

Equations for the calculation of short-circuit current

Note: Refer to TR50480 (currently a draft CENELEC report) for more information on this subject.

5.1 General equation for fault current

The general equation for fault current is:

$$I_f = \frac{V}{\sqrt{(R^2 + X^2)}} \qquad (5.1)$$

where:

I_f	is the fault current (amperes)
V	is the voltage between conductors at the supply point, either U_0 or U (volts)
R	is the resistance of the fault loop from source to point of fault (ohms)
X	is the reactance of the fault loop from source to point of fault (ohms).

The resistance and reactance are made up of a number of components, whose values depend on the type of fault and the supply system (see Figure 5.1).

Generally:

$$R = R_s + R_l + R_n$$

$$X = X_s + X_l + X_n$$

where the suffixes indicate:

- s supply or source quantity. Where power is obtained from a distributor these are external to the installation. It must be remembered that public electricity supply systems change to reflect growth or decline in electricity consumption for the local network and to meet day-to-day operational requirements. In general, a distributor cannot give a single unalterable value for the supply impedance.
- l line conductor quantity; that is, the total resistance of the line conductor(s) carrying the same fault current within the installation.
- n neutral conductor quantity, as for l.

In most cases there is a need to establish the minimum and maximum values and these can be obtained by reference to Electricity Association publications (now the Energy Networks Association), Engineering Recommendations P23/1, P25/1 and P26. For a user-owned source, a single value will generally be sufficient and this can be obtained from the manufacturer of the source equipment.

▼ Figure 5.1

General arrangement of impedances

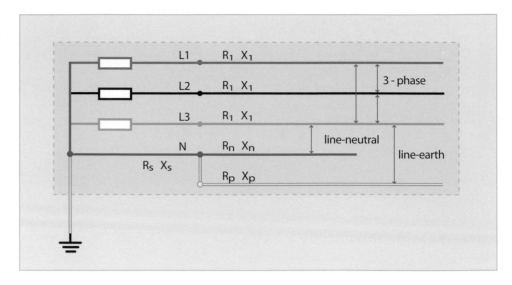

The terms R, X and Z are referred to as though they relate to a single length of conductor. It is, of course, clear that in practice they will relate to a number of conductors of different sizes in series carrying the same fault current and contributing towards the total values of R, X and Z. The value of a term in the equations given is the sum of the values for each portion of the total route from the supply point to the point of fault.

This convention is of importance for final circuits fed from distribution boards, because the upstream distribution circuits may, dependent on the lengths involved, contribute significant values of impedance.

5.2 Single-phase, line to neutral fault

The single-phase short-circuit current for a line to neutral fault is obtained from the following equations.

5.2.1 Circuits up to about 100 A

For single-phase supplies up to about 100 A:

i The separate values of R_s and X_s are not always available and an 'impedance' Z_s is substituted for R_s with $X_s = 0$. Depending on the length of the service cable the minimum value of Z_s will be in the range 0.015 to 0.15 ohm and the maximum value generally does not exceed 0.35 ohm, except in remote rural installations with low maximum demands, where Z_s could reach 0.5 ohm or more.

GN5

ii $(X_1 + X_n)$ can be neglected, so that

$$I_f = \frac{U_0}{Z_s + R_1 + R_n} \tag{5.2}$$

where:

U_0 is the line to neutral supply voltage
R_1 and R_n refer to the installation fault loop.

5.2.2 Circuits for more than 100 A

For circuits using conductor sizes of 35 mm² or larger ($X_1 + X_n$), R_s and X_s should be included in the calculation and the single-phase line to neutral short-circuit current is obtained by the following equation:

$$I_f = \frac{U_0}{\sqrt{(R_s + R_1 + R_n)^2 + (X_s + X_1 + X_n)^2}}$$ (5.3)

where:

R_s and X_s refer to a line-neutral supply loop
($R_1 + R_n$) and ($X_1 + X_n$) refer to the installation line-neutral loop.

5.2.3 Sources for values of resistance and reactance

R_1 and R_n can be calculated using conductor resistances from BS EN 60228 *Conductors of insulated cables*. Derivation of loop reactance ($X_1 + X_n$) can be by any standard method; suitable equations are provided in Appendix A of this Guidance Note.

BS EN 60228

Alternatively, combined values of ($R_1 + R_n$), and where applicable of ($X_1 + X_n$), can be derived from the tables for single-phase voltage drop for the appropriate two-core cables, or two single-core cables, in Appendix 4 of BS 7671.

Appx 4

Note that the values in the voltage drop tables are in effect expressed in impedance units of milliohms per metre at the conductor operating temperature (e.g. 70 °C for thermoplastic). For example, for use in the above equations:

$$(R_1 + R_n) = \frac{\text{Tabulated (r) x length}}{1000} \text{ ohms}$$

5.3 Conductor temperature and resistance

Resistance values taken from sources such as BS EN 60228 generally relate to a conductor temperature of 20 °C, whereas those from Appendix 4 of BS 7671 apply to the permissible maximum operating temperature of the particular type of cable. These resistance values may need adjustment to other temperatures, depending on the assumptions to be made for the particular fault current calculation.

There may be a degree of compromise necessary in deciding on the temperature at which the resistance of each conductor in the fault loop should be calculated. (It should be noted that circuit reactance is not temperature dependent and is not involved in these considerations.)

The following guidance on temperatures and resistances to be used for the calculation of fault current applies to the 'smallest' conductor(s) in a fault loop. Such a description includes a reduced section neutral, a protective conductor or the live conductors of the final circuit and assumes that all other conductors in the loop have equal or larger electrical cross-sectional areas. It is accurate enough for practical purposes, and is on the safe side, if all other conductors (i.e. upstream circuits) are assumed to have resistances corresponding to their working temperatures.

GN5

411.3.2

411.4.5

5.3.1 Shock protection

Regulation 411.3.2 is concerned with automatic disconnection in case of a fault and prescribes maximum times to be met. To achieve these limits, the loop impedance of circuits, Z_s, has to be limited. The equation of Regulation 411.4.5 is:

$$Z_s \leq \frac{U_0}{I_a}$$

where:

Z_s is the impedance in ohms (Ω) of the earth fault loop comprising:
- the source
- the line conductor up to the point of the fault, and
- the protective conductor between the point of the fault and the source

Table 41.1

I_a is the current in amperes (A) causing the automatic operation of the disconnecting device within the time specified in Table 41.1 of Regulation 411.3.2.2 or, as appropriate, Regulation 411.3.2.3 or 411.3.2.4. Where an RCD is used, this current is the rated residual operating current providing disconnection in the time specified in Table 41.1 or Regulation 411.3.2.3 or 411.3.2.4.

U_0 is the nominal a.c. rms or d.c. line voltage to Earth in volts (V).

The requirement is to ensure that Z_s is of a sufficiently low value that disconnection will occur within the required time.

Tables 41.2
to 41.4

The note below Tables 41.2, 41.3 and 41.4 reminds the user of the requirement that account must be taken of conductor temperature. The note advises that the loop impedances should not be exceeded when the conductors are at their normal operating temperature. This means that the impedances given in the tables cannot be used for testing at cold (ambient) temperatures, but must be reduced. For example, if testing 70 °C thermoplastic cables, the loop impedances in the tables must be divided by 1.2 (see column 3 of Table 5.1) to obtain test loop impedances at 20 °C ambient.

▼ **Table 5.1**
Resistance coefficients

		Multiplying coefficient		
Conductor operating temperature (°C)	Limiting final temperature (°C)	20 °C to operating temperature	20 °C to average of operating temperature and final temperature	Conductor working temperature to average of working temperature and final temperature
1	2	3	4	5
60	200	1.16	1.44	1.28
70	160	1.20	1.38	1.18
70	140	1.20	1.34	1.14
85	220	1.26	1.53	1.27
90	160	1.28	1.42	1.14
90	140	1.28	1.38	1.10
90	250	1.28	1.60	1.32

Quite clearly, reductions in supply voltage will increase disconnection times and the designer may wish to make appropriate allowances for this. Again, this is particularly important if specific device characteristics are being used, and not the worst-case ones of BS 7671.

5.3.2 Protection against short-circuit

Regulation 434.5.2 requires that where a protective device is provided for fault current protection only (and not overload protection), it shall be ensured that the disconnection time t is such that the live conductor temperature shall not exceed those temperatures given in Table 43.1. The formula quoted is:

434.5.2

Table 43.1

$$t = \frac{k^2 S^2}{I^2}$$

where:

    t   is the duration in seconds

    S   is the cross-sectional area of conductor in mm^2

    I   is the effective fault current, in amperes, expressed for a.c. as the rms value, due account being taken of the current-limiting effect of the circuit impedances

    k   is a factor taking account of the resistivity, temperature coefficient and heat capacity of the conductor material, and the appropriate initial and final temperatures. For common materials, the values of k are shown in Table 43.1.

This equation applies to the live conductors. The temperature corrections to be made will be as for shock protection in paragraph 5.3.1, i.e. to working temperature for devices in Appendix 3 of BS 7671 (column 3 of Table 5.1) and to average of working temperature and limiting final temperature for other devices.

Appx 3

5.3.3 Earth fault currents

Regulation 543.1.3 states that the cross-sectional area of protective conductors, where calculated, shall be not less than the value determined by the formula:

543.1.3

$$S = \frac{\sqrt{I^2 t}}{k}$$

In determining the current, account may need to be taken of the effect on the resistance of the circuit conductors of their temperature rise as a result of overcurrent.

Again, there is a presumption that it is appropriate to correct conductor resistance to the maximum operating temperature only.

Where reduced cross-section protective conductors are used, together with a 5 second disconnection time, the designer will need to take particular care as it is under these conditions that there is the most risk of the conductor overheating.

During a fault, each conductor increases in temperature. However, the conductor with the smallest cross-sectional area will experience the greatest increase and is the one to which the equation in Regulation 543.1.3 is applied. A reduced section neutral or protective conductor will rise in temperature more than a full section live conductor, and care needs to be taken.

543.1.3

In assessing earth fault current capabilities as well as the device characteristics used and the effect of load conditioning, the designer may wish to give consideration to other factors, including:

i   the supply (or external) loop impedance, Z_e
ii   the voltage tolerance.

If the supply impedance used in calculations is the maximum specified by electricity distributors, i.e. 0.8 Ω for TN-S and 0.35 Ω for TN-C-S supplies, as opposed to a measured value, there is a good probability of there being a safety factor implied.

Appx 2   The voltage range allowed to distributors is 230/400 V +10% −6%. If it is known that the voltage is low, allowance may need to be made, as reduced voltages increase disconnection times.

It is particularly important to take care with circuits where the nominal disconnection time is expected to be longer than 0.4 s, with a reduced section protective conductor, using a measured value of loop impedance and the possibility of a supply voltage at the lower end of the range. In these circumstances it could be prudent to include an additional margin of safety into the design by assuming a protective conductor resistance corresponding to the average of the appropriate assumed initial (working) temperature and final temperature given in Tables 54.2 to 54.6 in BS 7671. Suitable temperature adjustment factors (coefficients) are given in column 4 of Table 5.1 in paragraph 5.3.1.

Tables 54.2 to 54.6 *(margin note)*

5.4   Single-phase circuits

5.4.1   Single-phase circuits fed from three-phase board
Where a single-phase circuit is supplied from a three-phase and neutral board, the portions of $(R_1 + R_n)$ and $(X_1 + X_n)$ upstream of the board, if derived from the voltage drop tables of Appendix 4 of BS 7671, should be the values for single-phase circuits, i.e. for two-core cables (or two single-core cables if the upstream cables are single-core), having the same conductor size as the cable(s) supplying the board. They should not be taken from the values for three- or four-core cables or circuits.

Appx 4 *(margin note)*

If the upstream cable has a reduced neutral, the mean of the tabulated values of (r) and (x) for two-core cables of each conductor size should be used.

5.4.2   Effect of single-core cable spacing
Appx 4   The reactance values in Appendix 4 of BS 7671 assume that single-core cables are installed with an axial cable spacing of either one or two cable diameters (i.e. touching or a clearance of D_e). A larger cable spacing will increase the reactance. Figure 5.2 gives an approximate increase in loop reactance to be added to the term $(X_1 + X_n)$ for larger values of spacing of unarmoured single-core cables. For armoured cables reference should be made to Appendix A of this Guidance Note.

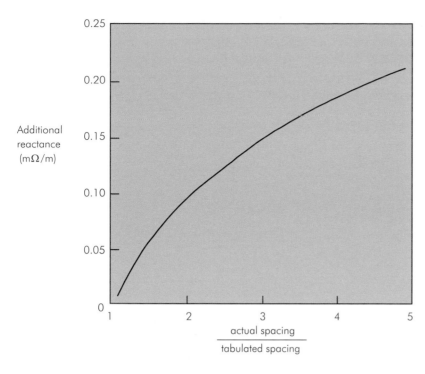

Additional
reactance
(mΩ/m)

0.25

0.20

0.15

0.10

0.05

0

1   2   3   4   5

$$\frac{\text{actual spacing}}{\text{tabulated spacing}}$$

5.5   Line-to-line short-circuit

A short-circuit may occur between two lines of a three-phase circuit, in which case the current is usually less than that of a three-phase fault. It may also occur with a single-phase line-to-line circuit where the neutral is not distributed.

5.5.1   Circuits up to about 100 A

For most practical purposes for circuits up to about 100 A, reactances X_s and X_1 can be neglected and R_s may be replaced by Z_s, where Z_s in this instance represents the line-to-line supply 'impedance'. R_n and X_n are not involved.

The short-circuit current is then:

$$I_f = \frac{U}{Z_s + 2R_1} \tag{5.4}$$

where:

U   is the line-to-line voltage
R_1 is the resistance of one line.

5.5.2   Circuits for more than 100 A

For circuits with conductors of 35 mm^2 or larger, the reactances X_s and X_1 should be included and equation (5.5) is used.

$$I_f = \frac{U}{\sqrt{(R_s + 2R_1)^2 + (X_s + 2X_1)^2}} \tag{5.5}$$

where:

R_s is the line-to-line resistance of the supply
X_s is the line-to-line reactance of the supply
R_1 and X_1 are for one line in the installation.

5.5.3 Sources for values of resistance and reactance

Appx 4

For multicore cable circuits, values of R_1 and X_1 can be obtained by calculation, or from Appendix 4 of BS 7671 by dividing the tabulated values for (r) and (x) for three-core cables by $\sqrt{3}$. Values of resistance may be adjusted to the appropriate conductor temperature using a coefficient from Table 5.1 given earlier.

For single-core cables, reference should be made to Appendix A of this Guidance Note, bearing in mind that for three cables in flat formation the lowest value of fault current occurs with a fault between the outer conductors.

5.6 Three-phase short-circuit

434.5.1
434.5.2
533.3

For the purpose of selection of a short-circuit protective device, for Regulations 434.5.1, 434.5.2 and 533.3, the three-phase short-circuit current is required. Distribution of a neutral does not affect the calculation.

5.6.1 Circuits up to about 100 A

X_1 is neglected, and R_s and X_s are replaced by Z_s, the line to neutral 'impedance' of the supply.

$$I_f = \frac{U/\sqrt{3}}{Z_s + R_1} \tag{5.6}$$

where:

R_1 is the resistance of one line
U is the line-to-line voltage.

5.6.2 Circuits for more than 100 A

For circuits with conductors of 35 mm^2 or larger, both X_s and X_1 should be included. The fault current is then obtained from equation (5.7).

$$I_f = \frac{U/\sqrt{3}}{\sqrt{(R_s + R_1)^2 + (X_s + X_1)^2}} \tag{5.7}$$

where :

R_s is the line-neutral resistance of the supply
X_s is the line-neutral reactance of the supply
R_1 and X_1 are for one line in the installation
U is the line-to-line voltage.

5.6.3 Sources for values of resistance and reactance

The resistance R_1 can be calculated by reference to BS EN 60228 (see section 5.2.3). Appendix A of this Guidance Note provides information on the derivation of reactance X_1.

Appx 4

With the exception of fault loops whose impedance is attributable mainly to large single-core cables in flat formation, values of R_1 and X_1 can also be obtained from the voltage drop tables in Appendix 4 of BS 7671 for the appropriate three-phase installation.

Note that the three-phase tabulated values of (r) and (x) in Appendix 4 are, in fact, $\sqrt{3}R_1$ and $\sqrt{3}X_1$, so that, in addition to the correction of (r) for temperature, both tabulated values must be divided by $\sqrt{3}$.

Corrections to R_1 for temperature (i.e. for the highest and lowest prospective currents) apply as in the single-phase case, and the same table of coefficients can be used (Table 5.1).

5.6.4 Special considerations with single-core cables

Where all of the cabling from supply to fault is multicore type, or single-core type installed in trefoil, the reactance X_1 will be the same for all lines and the fault currents will be balanced.

If a minor portion of the fault loop impedance is attributable to single-core cables in flat formation, an average value for X_1 is appropriate. This is obtained by deriving X_1 for a conductor spacing equal to 1.26 s, where s is the axial spacing between adjacent cables; see Appendix A or Figure 5.2 of this Guidance Note.

Where a substantial part of the impedance from supply to fault is contributed by a large single-core type of cable arranged in flat formation, the value of reactance for the centre cable should be used. Methods for computing this reactance are given in sections A.4.3 and A.4.4 of Appendix A of this Guidance Note.

Equations for the calculation of earth fault current

6

Note: Refer to TR50480 (currently a draft CENELEC report) for more information on this subject.

6.1 General

Earth fault current has to be calculated for:

1 the selection of the size of the circuit protective conductor (cpc) for its limiting temperature rise   543.1.3

2 checking the selection of the line conductor (in both single-phase and three-phase circuits), because the earth fault loop may have a lower impedance than the short-circuit loop   434.1

3 checking the breaking capacity of the device intended to interrupt earth fault current

4 use in connection with meeting the shock protection requirements of Chapter 41 of BS 7671.   Chap 41

6.2 TN systems (TN-C, TN-S and TN-C-S)

The particular feature of these systems is that there is a continuous metallic path from the exposed metalwork (exposed-conductive-parts) of equipment to the neutral of the source (see Figure 6.1). For such a path the value of overcurrent is high enough to operate protection such as a fuse or circuit-breaker.   Fig 2.2 Fig 2.3 Fig 2.4

For reasons explained earlier, distributors cannot give a single definitive value for Z_s, but have published guidance on the maximum values, which can be used to calculate minimum disconnection times, in Engineering Recommendation P23/1 *Consumer's earth fault protection for compliance with the IEE Wiring Regulations for Electrical Installations* (originally published by the Electricity Associations and now available from the Energy Networks Association). Alternatively, a value of $(R_s + X_s)$ for the line-earth fault path of the supply for supplies up to about 100 A may be measured between terminals A and 0 with the installation disconnected. In this latter case, due consideration must be given to the possibility that modifications in the supply network may change the impedance, so consultation with the distributor is usually desirable. For user-owned sources, the information should be obtained from the equipment manufacturer.

▼ **Figure 6.1**
Earth fault path, TN-C-S
system

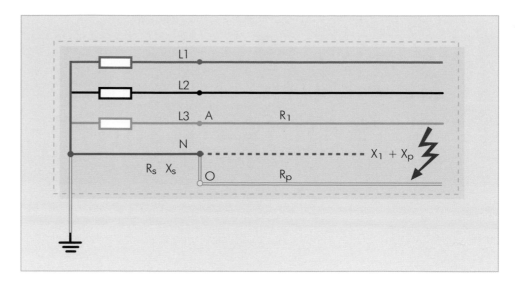

6.2.1 Circuits up to about 100 A

For circuits or supplies up to about 100 A, X_s is neglected and only R_s is used, although it may be referred to as an 'impedance'.

For such supplies, typical maximum values of R_s are about 0.35 ohm if the supply is TN-C-S, or about 0.8 ohm if it is TN-S.

$$I_f = \frac{U_0}{R_s + R_1 + R_p} \tag{6.1}$$

where:

  R_1 is the line resistance
  R_p is the protective conductor resistance
  $(X_1 + X_p)$ is assumed to be zero.

This assumption depends on the two conductors being installed close to each other, which is good practice, and there being no ferromagnetic material between them. It follows that a separate protective conductor, if required, should be installed together with the appropriate line and neutral conductors inside any steel enclosure.

521.8.1

6.2.2 Circuits for more than 100 A

For circuits or supplies for more than 100 A the earth fault current is calculated from equation (6.2):

$$I_f = \frac{U_0}{\sqrt{(R_s + R_1 + R_p)^2 + (X_s + X_1 + X_p)^2}} \tag{6.2}$$

where $X_1 + X_p$ = reactance of the line-protective conductor loop.

Generally applicable values for three-phase supplies up to 300 A are given in Engineering Recommendation P23/1. Alternatively, the distributor or the manufacturer of user-owned sources should be consulted for values of R_s and X_s.

6.2.3　Sources for values of resistance and reactance

R_1 can be obtained from Appendix 9 of the *On-Site Guide* or Appendix E of Guidance Note 1 (at 20 °C), or from the voltage drop tables in Appendix 4 of BS 7671 (at conductor operating temperature). For the purposes of Regulation 434.5.2 the value of R_1 has to be corrected for temperature; see section 5.3 on resistances for short-circuit calculations. The same coefficients can be used.

OSG, Appx 9
GN1, Appx E
Appx 4
434.5.2

If the voltage drop tables of BS 7671 are used, the tabulated resistance values have to be divided by two in the case of values for two-core cables and by $\sqrt{3}$ for three- and four-core cables, because only one live conductor is involved.

Where the protective conductor consists of a standard size conductor, the value of R_p can be obtained from the same sources as for a line conductor but using temperature adjustments appropriate to the type of protective conductor involved (see Tables 54.2 to 54.5 in BS 7671 and Table 5.1 of this Guidance Note).

It may sometimes be possible, where the protective conductor is reasonably close to the line conductor, to avoid computation of $(X_1 + X_p)$ by adapting the reactance for the line-neutral loop for unarmoured single-phase single-core cable in the voltage drop tables in Appendix 4 of BS 7671. Where the protective conductor is a different size from the line conductor, the average reactance of the two conductor sizes should be used. Corrections will probably be required for a wider spacing. The procedures described for single-phase short-circuits can be followed.

Appx 4

For armoured cables and cables in conduit and trunking, where the enclosure is used as the protective conductor, special values of the loop parameters are provided below.

6.2.4　Value of touch voltage

The impedance across which a touch voltage appears is:

$$\sqrt{R_p{}^2 + X_p{}^2} \tag{6.3}$$

Where X_p has not been determined separately it can, for convenience, be assumed to be $(X_1 + X_p)/2$.

Where the size of the protective conductor is 35 mm² or less, X_p may be neglected.

6.3　Use of a cable enclosure as a protective conductor

Where the sheath or armour of a cable or metallic conduit or trunking is used as a protective conductor, R_p and $(X_1 + X_p)$ can be derived from the following data.

6.3.1　Steel-wire armoured multicore cables

$$(R_1 + R_p) = R_c + 1.1\,R_a \tag{6.4}$$

$$(X_1 + X_p) = \frac{0.3 \times L}{1000}\,\Omega \tag{6.5}$$

where:

　L　is the conductor length in metres
　R_c　is the resistance of the line conductor

R_a is the d.c. resistance of the armour adjusted, as necessary, to the appropriate temperature. Values of armour resistance at 20 °C are given in the appropriate cable standard

1.1 is a coefficient which represents the magnetic effect of the steel armour.

The behaviour of steel-wire armour as a protective conductor and its temperature rise with fault current is not well understood. The problem can be simplified by considering that the steel wire behaves like a copper conductor having roughly one-half of the cross-sectional area of the steel. For small cables the armour cross-sectional area is more than adequate, while for the largest sizes its performance is nearer to that of a copper protective conductor having a cross-sectional area about 50 per cent of that of the line conductors.

Selection of the appropriate resistance adjustment for temperature when calculating earth fault loop impedances with steel-wire armoured cables follows Table 5.1 of this Guidance Note for copper protective conductors, but with the steel armour credited with about one-half of its cross-sectional area. Armour cross-sectional areas can be obtained from BS 6346 or BS 5467.

Suitable temperature adjustment coefficients for the resistance of steel-wire armour, to be used as multipliers to 20 °C values (see BS 6346 or BS 5467), are given in Table 6.1.

▼ **Table 6.1**
Coefficients for steel-wire armour (to be applied to resistances at 20 °C)

Insulation material	Thermoplastic		Thermosetting
	70 °C	**90 °C**	**90 °C**
Assumed initial temperature (operating)	60 °C	80 °C	80 °C
Final temperature	200 °C	200 °C	200 °C
Coefficient for adjustment to operating temperature	1.18	1.27	1.27
Coefficient for adjustment to the average of the initial and final temperatures	1.50	1.54	1.54

It is important to note here that, because the line conductor is contained within the armour, the latter acting as the return conductor, the reactance $(X_1 + X_p)$ is practically entirely associated with the line conductor.

The steel-wire armour on multicore cables is usually quite adequate for earth fault current duty but, in the unlikely event that this is not so, it is better to select a larger size of cable rather than to connect an auxiliary conductor in parallel with the armour. The magnetic effect of the armour will restrict the current passing through such a conductor to a small fraction of that expected on the basis of division according to d.c. resistances (see section 6.3.4).

Touch voltage
The impedance for the calculation of touch voltage is R_a and there is no reactive term.

6.3.2 Aluminium-wire armoured single-core cables

The armour of these cables is usually bonded together to form an earth fault return through the three armour paths in parallel. Because these paths are of low resistance, inductive effects influence the division of current between them and, except for the smaller sizes of cable, the parallel resistance is greater than $R_a/3$. Further, there is a small, but in some cases significant, reactive component to the armour impedance.

$$(R_1 + R_p) = R_c + C_r R_a \qquad\qquad (6.6)$$

$$(X_1 + X_p) = X_1 + C_x R_a$$

where:

R_c is the resistance of conductor at the appropriate temperature

R_a is the d.c. resistance of armour of one cable at the appropriate temperature (see Table 5.1 of this Guidance Note)

X_1 is the reactance of the conductor armour loop, or the internal reactance of a single cable. A method for calculating X_1 for an isolated cable is given in paragraph A.5 of Appendix A of this Guidance Note

C_r is a coefficient to be applied as a multiplier to R_a to obtain the effective resistance of three armours in parallel. The value of C_r depends on the size and arrangement of the cables

C_x is a coefficient to be applied as a multiplier to R_a to obtain the effective reactance of three armours in parallel. The value of C_x depends on the size and arrangement of the cables.

Values of C_r and C_x are given in Table 6.2 for common arrangements of single-core cables to BS 6346 and BS 5467.

| Size (mm²)[3] | Trefoil | | 3 flat formation[2] | | | |
| | | | Touching | | Spaced[4] | |
	C_r	C_x	C_r	C_x	C_r	C_x
150	0.35	0.09	0.35	0.14	0.38	0.20
185	0.35	0.10	0.35	0.15	0.39	0.21
240	0.35	0.11	0.36	0.16	0.40	0.23
300	0.35	0.12	0.36	0.17	0.41	0.25
400	0.37	0.16	0.39	0.24	0.47	0.31
500	0.38	0.17	0.39	0.25	0.49	0.33
630	0.39	0.18	0.40	0.26	0.51	0.33
800	0.43	0.23	0.45	0.33	0.60	0.37
1000	0.44	0.24	0.47	0.34	0.63	0.37

▼ Table 6.2
Values of C_r and C_x for aluminium-wire[1] armoured single-core copper conductor cables

Notes:

1 Values for solid aluminium conductor cables are approximately the same as those for copper conductor cables having the same diameter.

2 Earth fault assumed to be between an outer conductor and bonded armour. Values for a fault from the central conductor are slightly lower.

3 For sizes up to and including 120 mm², C_r has the conventional value of 1/3 and $C_x = 0$.

4 Axial spacing 2 x D_e.

The above relates to a circuit having one cable per phase with the fault at the far end of the run. Where a fault along the run is envisaged, or there is more than one cable per phase, special considerations are necessary (see Regulation 434.4).

434.4

543.1.3 Resistance of armour at 20 °C is available from the cable standard and can be corrected to the appropriate temperature. For calculating the highest value of earth fault current a resistance at 20 °C is suitable. When checking compliance with Regulation 543.1.3, a resistance adjustment is required (see Table 5.1).

Suitable coefficients, to be applied as multipliers to 20 °C values, are given in Table 6.3.

▼ **Table 6.3**
Coefficients for aluminium-wire armour

| Insulation material | Thermoplastic | | Thermosetting |
	70 °C	90 °C	90 °C
Assumed initial temperature (operating)	60 °C	80 °C	80 °C
Final temperature	200 °C	200 °C	200 °C
Coefficient for adjustment from 20 °C to operating temperature	1.16	1.24	1.24
Coefficient for adjustment from 20 °C to the average of the operating and final temperatures	1.44	1.48	1.48

Touch voltage

For the smaller sizes of cable, the touch voltage is calculated using R/3, the d.c. resistance of the three armours in parallel, and the reactive term can be ignored. For larger sizes, the impedance responsible for the touch voltage is

$$\sqrt{(C_r\,R_a)^2 + (C_x\,R_a)^2}$$

6.3.3 Copper sheathed cables

The approach is similar to that for aluminium-wire armoured cables, with the following modifications:

X_1 and X_p are both negligible.

For multicore cables, R_p is the sheath resistance adjusted to the appropriate temperature. For single-core cables, R_p must take account of the number of sheaths bonded together.

Resistances for conductor (R_1) and sheath (R_p) at 20 °C can be obtained from data given in BS EN 60702-1, or from the manufacturer, and adjusted to the appropriate temperatures.

To adjust resistances at 20 °C to the highest average temperature permitted during an earth fault, the coefficients in Table 6.4 may be used.

▼ **Table 6.4**
Coefficients for copper sheaths

Assumed initial temperature	70 °C	105 °C
Final temperature	200 °C	200 °C
Coefficient for adjustment from 20 °C to operating temperature	1.20	1.33
Coefficient for adjustment from 20 °C to the average of the operating and final temperatures	1.45	1.52
543.1.3 **Value of k (Regulation 543.1.3)**	1.35	1.12

Touch voltage

The impedance for the calculation of touch voltage is R_s, the sheath d.c. resistance, and there is no reactive term.

6.3.4 Auxiliary conductors

The use of an auxiliary conductor to supplement the fault current capacity of aluminium-wire armour or metallic sheaths calls for the placement of such a conductor as close as possible to the cable. Even when such a conductor is touching the cable, its current is less than that indicated by the respective d.c. resistances. When the separation is of the order of 100 mm or more, the reactive effect will reduce the effectiveness of the auxiliary conductor considerably. (The interposition of ferromagnetic material would decrease the effectiveness dramatically.)

When cables are in flat formation, the use of two auxiliary conductors placed between the cables is a way of achieving a small separation from all lines.

6.3.5 Cables in steel conduit

The magnetic effect of steel conduit is quite significant and affects the loop resistance and reactance. Because the effect is non-linear with current, two ranges of fault current have to be recognised in order to avoid making the design procedure unreasonably complicated. Accuracy for values of effective resistance and reactance is poor because of the variability of the magnetic properties of steel.

The effective (line + protective conductor) resistance is:

$$(R_1 + R_p) = R_c + F_r R_{dc} \tag{6.7}$$

where:

R_c is the resistance of the line conductor
R_{dc} is the d.c. resistance of the conduit at working temperature (assumed to be about 50 °C, so the adjustment to be applied to values at 20 °C is to multiply by 1.12)
F_r is a factor to take account of the magnetic effect of the steel.

The effective reactance of the line-protective conductor loop is given by:

$$(X_1 + X_p) = F_x R_{dc} \tag{6.8}$$

where F_x is a factor to take account of the reactive effect of the steel.

The unexpected feature of equation (6.8), where the reactance is stated to be dependent on conduit resistance, comes about because the major contributor to reactance of the conductor/conduit loop is the magnetic flux within the steel. The distribution of this flux within the wall of the conduit and its value depends on the frequency of the current and the resistivity and permeability of the steel. The factor F_x is empirical, being derived from tests at 50 Hz on samples of conduit made to BS 4568 (now superseded by BS EN 61386). Because of the variable magnetic properties of conduit steel, the values given for F_x (and F_r) should be regarded as typical.

Values of F_r and F_x are given in Table 6.5 for both heavy and light gauge steel conduit.

Heavy gauge conduit

Size of conduit (mm)	Fault current up to 100 A		Fault current above 100 A	
	F_r	F_x	F_r	F_x
16	3.0	2.0	1.3	1.3
20	2.8	1.9	1.3	1.3
25	2.4	1.7	1.1	1.1
32	2.0	1.4	0.92	0.92

Light gauge conduit

Size of conduit (mm)	Fault current up to 100 A		Fault current above 100 A	
	F_r	F_x	F_r	F_x
16	2.3	1.6	1.3	1.3
20	2.1	1.4	1.3	1.3
25	1.9	1.3	1.1	1.1
32	1.8	1.3	1.1	1.1

Touch voltage

The apparent impedance of the conduit is, for usual values of fault current, much less than its d.c. resistance. This is due to the magnetic screening effect of steel. However, for simplicity and to be on the safe side when calculating a touch voltage, it is assumed to be equal to its resistance, R_{dc}, irrespective of the value of current. There is no reactive component.

6.3.6   Cables in steel ducts and trunking

The magnetic effect of steel trunking is rather less than that of conduit, but the method for deriving values of resistance and reactance is similar. Trunking is not suitable for use as a protective conductor for circuits carrying much more than 100 A; unless particular care is taken to ensure the continuity and current-carrying capability of joints, it is possible to use only one range of fault current.

For circuits carrying no more than 100 A it is possible to derive values for $(R_1 + R_p)$ and $(X_1 + X_p)$ by one set of formulae:

$$(R_1 + R_p) = R_c + 2.1 \, R_{dc} \tag{6.9}$$

where:

    R_c is the resistance of the line conductor
    R_{dc} is the d.c. resistance of the trunking at 50 °C
    2.1 is a coefficient to take account of the magnetic effect of the steel.

$$(X_1 + X_p) = 2 \, R_{dc} \tag{6.10}$$

Touch voltage

The impedance to be used to calculate touch voltage is R_{dc}, with no reactive term.

No protective conductor or auxiliary protective conductor should be placed outside steel conduit or trunking containing the associated live conductors. The magnetic screening effect of the conduit or trunking is so great that any external conductor will be of little benefit.

6.4   TT system

The feature of this system is that there is no continuous metallic path between the exposed metalwork (exposed-conductive-parts) of the equipment and the neutral of the source (see Figure 6.2). Part of the return path goes via two earth electrodes and the general mass of Earth. The resistance of the electrodes is usually greater than 1 ohm, perhaps up to 200 ohms for the installation electrode, above which it is likely to be unstable.

Guidance on the installation and testing of these electrodes is given in BS 7430:1998 *Code of Practice for Earthing*.

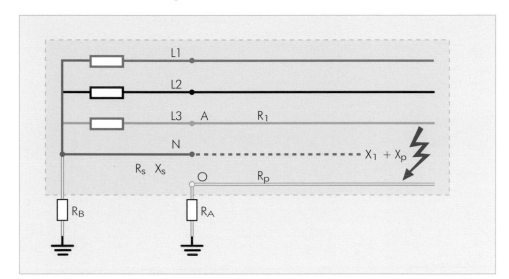

▼ **Figure 6.2**
Earth fault path,
TT system

Due to the earth return path via the electrodes, earth fault currents are generally much lower than for TN systems. For this reason it is usually not possible to use automatic disconnection by an overcurrent device as a means of shock protection.   411.5.2

Calculation of earth fault current follows the same lines as those set out for TN systems, but the resistances of the earth electrodes, R_A and R_B, have to be added to the external portion of the earth fault loop (see Figure 6.2). Values of R_A and R_B are generally much greater than those of R_s and X_s experienced with TN systems and almost always have to be measured or estimated for each individual installation.

The earth fault current is given by:

$$I_f = \frac{U_0}{R_B + R_A + (R_1 + R_p)} \qquad (6.11)$$

where:

R_B is the earth electrode resistance at source
R_A is the resistance of the installation earth electrode.

In most installations likely to use this form of earthing with automatic overcurrent protection to comply with Regulation 411.5.2, $(R_1 + R_p)$ can usually be neglected in   411.5.2
comparison with $(R_A + R_B)$.

Further, the usual value of $(R_A + R_B)$ is such that all reactances can be ignored.

Touch voltage

The impedance across which a touch voltage is produced is ($R_A + R_p$) but, as noted above, R_p can usually be neglected.

Selection of conductor size

7

7.1 General

Conductor sizes selected to meet the requirements for current-carrying capacity, voltage drop and, in the case of protective conductors, the requirements of Section 411, have to be checked to confirm that they will not be overheated in the event of a fault.

Section 523
Section 525
Section 411

For this purpose the following regulations apply:

▶ Short-circuit, line and neutral	434.5.2	*434.5.2*
▶ Earth fault, line	434.5.2	
▶ Earth fault, protective conductor	543.1.3	*543.1.3*

7.2 Overload and short-circuit protection by the same device

Regulation 434.5.1 allows a designer to assume that if an overcurrent protective device is providing overload protection complying with Section 433, and has a rated breaking capacity not less than the prospective short-circuit current at its point of installation, it is also providing, without further proof, short-circuit protection. That is to say, there is no need to check that the circuit conductors are adequately protected thermally by verifying compliance with Regulation 434.5.2.

434.5.1

434.5.2

This assumption holds true for all fuses and energy limiting circuit-breakers. For zero-point or current-zero type circuit-breakers there may be a value of fault current above which the conductors are not protected. It will generally be found in practice, however, that this value is greater than the prospective short-circuit current. Advice on the type of circuit-breaker should be sought from the manufacturer (see section 3.6 of this Guidance Note).

It is necessary to comply with Regulation 434.5.2 where short-circuit protection is provided by a separate device, and to check compliance where the characteristic of the proposed device for combined overload and fault current protection is in doubt.

7.3 Earth fault current

Although the size of a line conductor may comply with Regulation 434.5.2 under short-circuit conditions, where earth fault current is less than the short-circuit current the energy let-through of the device may be greater and it is necessary to check the sizes of both the line conductor for compliance with Regulation 434.5.2 and the protective conductor for compliance with Regulation 543.1.3.

434.5.2
543.1.1

543.1.3

7.4 Parallel cables

434.4 The assumption of Regulation 434.5.1 should not be made for conductors in parallel. Regulation 434.4 sets out the requirements for the protection of cables in parallel:

> A single protective device may protect conductors in parallel against the effects of fault currents provided that the operating characteristic of the device results in its effective operation should a fault occur at the most onerous position in one of the parallel conductors. Account shall be taken of the sharing of the fault currents between the parallel conductors. A fault can be fed from both ends of a parallel conductor.
>
> If the operation of a single protective device may not be effective then one or more of the following measures shall be taken:
>
> (i) The wiring is installed in such a manner as to reduce to a minimum the risk of a fault in any parallel conductor, for example, by the provision of protection against mechanical damage
> (ii) For two conductors in parallel, a fault current protective device is provided at the supply end of each parallel conductor
> (iii) For more than two conductors in parallel, fault current protective devices are provided at the supply and load ends of each parallel conductor.

7.5 Use of I²t characteristics

434.5.2
Appx 3

For short durations, say less than 0.1 s, the values derived from Appendix 3 of BS 7671 may be unsuitable because of the possible current-limiting effect of the device. The manufacturer should be consulted as to the I²t let-through for the expected value of prospective current.

It is then possible to substitute this value of I²t in a rearrangement of the equation of Regulation 434.5.2, or as given in Regulation 543.1.3, i.e.

543.1.3

$$S = \frac{\sqrt{I^2 t}}{k} \ \text{mm}^2$$

Table 43.1
Tables 54.2 to 54.6

to obtain the smallest conductor size to comply with the maximum permitted temperature in Table 43.1 or Tables 54.2 to 54.6 of BS 7671.

7.6 Duration of short-circuit current

434.5.2
543.1.3

Note that there is no limitation on fault current duration inherent in the thermal requirements of either Regulation 434.5.2 or Regulation 543.1.3. While considerations of shock risk limit earth fault currents to no more than 5 s, there is no definitive limit for short-circuit currents.

The duration for which a short-circuit current should be allowed to persist is a matter of careful judgement (see section 1.6 of this Guidance Note). Obviously, it is wise to restrict the duration of arcing, etc. to an absolute minimum. However, there is no firm evidence on which to base a specific time limit for general use, and consideration of this matter must take into account the particular circumstances.

The data on which the values of the coefficient k in Tables 43.1 and 54.2 to 54.6 of BS 7671 have been based were generally obtained with durations not greater than about 5 s in mind, but again, this is unlikely to be critical. It is nevertheless prudent to consult the cable manufacturer if a duration in excess of 5 s is contemplated.

Table 43.1
Tables 54.2 to 54.6

7.7   Status of adiabatic equations

The adiabatic equations contained in Regulations 434.5.2 and 543.1.3 provide an approximate value for duration or conductor size. It should be noted straightaway that any error is on the safe side, but may be of economic interest.

434.5.2
543.1.3

The validity of the adiabatic equation depends on the extent to which heat is lost from the conductor during the period of the fault current. For a given conductor type and insulation material, the heat loss can be expressed in terms of the ratio of fault current duration to the conductor cross-sectional area.

For circular conductors the limitation on accuracy can be expressed by the following examples.

For durations (in seconds) up to 0.1 x cross-sectional area (in mm²) the loss is negligible. In practical terms this means that the adiabatic equation is of sufficient accuracy at durations up to 0.4 s for conductors of 4 mm² or larger, up to 1 s for conductors of at least 10 mm² and up to 5 s for conductors of 50 mm² or larger.

For most fault current interruption times, the approximate nature of the adiabatic equation for circular conductors is of interest for the small conductor sizes only. Generally the economics for circuits of such sizes means that the more complicated non-adiabatic calculation is unlikely to be worthwhile.

Shaped conductors are in the size range where the approximation does not matter.

The situation changes considerably for concentric conductors such as metallic sheaths, armour and concentric neutrals. BS 7454 provides equations and data for making non-adiabatic calculations for all types of conductor and insulation material.

BS 7454

For bare conductors the adiabatic equation is satisfactory.

7.8   Alternative values of k

Note 2 to Regulation 434.5.2 states that other values of k can be determined by reference to BS 7454, such as where the initial temperature is lower than that assumed for Table 43.1. This situation arises when conductor sizes are increased above that required to comply with current-carrying capacities.

434.5.2

Such a recalculation will yield a higher value of k and permit slightly higher fault currents for a given size of conductor; the average conductor temperature should also be recalculated. The reduction in initial temperature has to be considerable for the recalculation to be worthwhile.

Appendix B of this Guidance Note provides the formula to calculate k.

7.9 Appendix 4 of BS 7671

Appx 4 Appendix 4 in BS 7671 has been amended to closely align with amended international standards. This has meant that the appendix gives additional installation methods, reference methods, rating factors and current-carrying capacities.

Table 4A2 The number of installation methods described in Table 4A2 (previously 4A1) has risen from 20 to 50-odd. Although this may appear to make things more complicated, the appendix now embraces installation methods that are used but which were not previously accounted for, including installation methods in building voids, direct in ground, in ducts in the ground, and flat twin and earth cables in thermal insulation.

It is impractical to calculate and publish current ratings for every installation method, since many would result in the same current rating. Therefore a suitable (limited) number of current ratings have been calculated which cover all of the installations stated in the Wiring Regulations, and are called reference methods.

All the individually numbered installation methods have a lettered reference method stated against them in Table 4A2, except for flat twin and earth cables which have reference method numbers 100 to 103. There are seven alphabetically lettered reference methods, that is A to G.

Appx 4
para 7

The lettered reference methods broadly cover the following areas:

- ▶ **Reference method A –** Enclosed in conduit in thermally insulated walls etc. (Note: The wall consists of an outer weatherproof skin, thermal insulation and an inner skin of wood or wood-like material having a thermal conductance of at least 10 W/m^{2}K. The conduit is fixed so as to be close to, but not necessarily touching, the inner skin. Heat from the cables is assumed to escape through the inner skin only. The conduit can be metal or plastic.)
- ▶ **Reference method B –** Enclosed in conduit on a wall or in trunking etc.
- ▶ **Reference method C –** Clipped direct.
- ▶ **Reference method D –** Direct in the ground or in ducting in the ground.
- ▶ **Reference method E –** Multicore cables in free air or on perforated trays etc.
- ▶ **Reference method F –** Single-core cable touching in free air or on perforated trays etc.
- ▶ **Reference method G –** Single-core cable spaced in free air or on perforated trays etc.

7.9.1 Comparison between old (16th Edition) and new reference methods

Old	New	Comments
1	C	No change in actual method
3	B	No change in actual method
4	A	No change in actual method
11 or 13	E	Reference methods 11 and 13 covered both single-core and multicore cables in free air, on ladder rack, and on perforated cable tray etc. Reference method E only covers multicore cables in free air, on ladder rack, and on perforated cable tray etc. For single-core cables see reference methods F and G.
12	F & G	Reference method 12 covered single-core cables in free air and on ladder rack. Reference method F covers single-core cables touching in free air, on ladder rack, and on perforated cable tray. Reference method G covers single-core cables spaced in free air, on ladder rack, and on perforated cable tray.
–	D	New method covering multicore armoured cables direct in ground and in ducts in ground.
70 °C thermoplastic insulated and sheathed flat cables with protective conductor:		
6	A	No change in actual method
15	100	See note 1
	101	New method, see note 2
	102	New method, see note 3
	103	New method, see note 4

Notes:

1 Reference method 100: cables installed above a plasterboard ceiling covered by thermal insulation **not exceeding 100 mm** in thickness.

2 Reference method 101: cables installed above a plasterboard ceiling covered by thermal insulation **exceeding 100 mm** in thickness.

3 Reference method 102: cables installed in a stud wall with thermal insulation and with the cable **touching the inner wall surface**.

4 Reference method 103: cables installed in a stud wall with thermal insulation and with the cable **not touching the inner wall surface**.

As can be seen from the above descriptions, apart from the addition of method D (cables in ground), and extending the current rating methods for flat twin and earth cable, little has changed with regard to current ratings.

Calculation of reactance

A

A.1 General

It is important to bear in mind that reactance is essentially associated with a circuit or loop, generally formed by two conductors. For this reason, the reactance of line-neutral and line-protective conductor loops has been expressed in this Guidance Note in the form $(X_1 + X_n)$ etc. so as to avoid the impression that the contribution by each conductor can be applied in an equation, measured or considered to exist, as an independent quantity.

However, because the two conductors may be of different size or form, it is convenient for the purpose of computation to regard their reactances separately, as one does the resistances. In the case of a three-phase fault, the return conductor is formed by the other lines and does not appear directly in the computation.

A.2 Line to neutral, single-phase, faults

A.2.1
Line and neutral reactances for:

i   all multicore cables
ii   unarmoured or non-metallic sheathed single-core cables.

The reactance of each conductor is given by:

$$X_1 \text{ and } X_n = 0.0628 \log_e(2\,s/a.d) \quad \text{m}\Omega/\text{m at 50 Hz}$$

where:

   s   is the axial separation of the two conductors (mm)
   d   is the conductor overall diameter (mm). Shaped conductors are treated as circular conductors having the same cross-sectional area
   a   is a coefficient to allow for the internal reactance of the conductor, according to Table A.1.

▼ Table A.1
Coefficient a for
calculation of reactance

No. wires in conductor	Coefficient, a
3	0.678
7	0.724
19	0.758
37	0.768
61	0.772
91	0.774
127	0.776
169	0.779

The reactance due to the helical laying up of cores of multicore cables can be neglected.

A.2.2

Resistance and reactance of line and neutral conductors for non-magnetic armoured or metal sheathed single-core cables.

(If the sheaths or armour are not bonded at both ends of the run, the impedance for non-metallic sheaths or unarmoured cables applies.)

In this case both resistance and reactance are modified by the presence of circulating current in the armour or sheath. The apparent values of resistance and reactance of each of the conductors are given by:

$$R_1 \text{ and } R_n = R_c + F\,R_a$$

$$X_1 \text{ and } X_n = X_c - F\,X_a$$

where:

R_c is the conductor resistance at the appropriate temperature
X_c is the reactance of each conductor, calculated as in section A.2.1
R_a is the resistance of the armour or sheath
X_a is the reactance of the sheath or armour
     $= 0.0628\,\log_e(2\,s/d_a)$ mΩ/m at 50 Hz
     where:
         d_a is the mean diameter of the sheath or armour (mm)
         s is the axial separation of the cables (mm)

F   is the coefficient of coupling between the armour and conductor, given by

$$F = \frac{X_a^2}{R_a^2 + X_a^2}$$

A.3 Line-to-line, single-phase, short-circuit

A.3.1

Reactance of each line conductor for:

i multicore cable
ii non-metal sheathed or unarmoured single-core cables in trefoil
iii adjacent non-metal sheathed or unarmoured single-core cables in flat formation.

X_1 is given by the same equation as for line to neutral faults in section A.2.1, with s being the axial separation of adjacent line conductors.

A.3.2

Resistance and reactance for each line conductor for:

i non-magnetic armoured or metal sheathed single-core cables in trefoil
ii adjacent non-magnetic armoured or metal sheathed single-core cables in flat formation.

The increased resistance and reduced reactance of each line conductor are given by the equations for R_1 and X_1 in section A.2.2, with s being the axial separation between adjacent line conductors or cables.

A.3.3

For a fault between the outer conductors of three single-core cables of any type in flat formation, equally spaced.

The reactance and, where applicable, the resistance of each line are given by the equations in sections A.2.1 and A.2.2, according to the type of cable and installation, but with the values for both X_c and X_a increased by 0.04355 mΩ/m. In these equations the symbol s is unchanged as the axial separation between adjacent conductors or cables.

A.4 Three-phase short-circuit

A.4.1

Reactance of each line conductor for:

i multicore cables
ii non-metal sheathed or unarmoured single-core cables in trefoil.

$$X_1 = 0.0628 \log_e(2 \, s/a.d) \quad \text{mΩ/m at 50 Hz}$$

where s is the axial separation between the line conductors (mm).

Other symbols and values of a are as before.

A.4.2

Resistance and reactance for non-magnetic armoured or metal sheathed single-core cables in trefoil.

The apparent increase in resistance R_1 and decrease in reactance X_1 of the line conductor are given by the equations in section A.2.2 with s being the axial separation between line conductors or cables.

A.4.3

Resistance and reactance for three unarmoured or non-metal sheathed single-core cables in flat formation, equally spaced.

The fault currents in the lines are not equal and the centre conductor has the lowest apparent reactance. For this conductor:

$$R_1 = R_c$$

$$X_1 = X_c - X_m/3$$

where:

   R_c is the conductor resistance at the appropriate temperature
   $X_c = 0.0628 \log_e(2 \, s/a.d)$ mΩ/m at 50 Hz
   $X_m = 0.0628 \log_e(2) = 0.04355$ mΩ/m at 50 Hz
   s   is the axial separation between adjacent line conductors (mm).

A.4.4

Resistance and reactance for three non-magnetic armoured or metal sheathed cables in flat formation, equally spaced.

The apparent increase in resistance and decrease in reactance of the middle conductor are given by the following equations:

$$R_1 = R_c + F \, R_a$$

$$X_1 = X_c - X_m/3 - F \, (X_a - X_m/3)$$

where:

   R_c is the resistance of the conductor at the appropriate temperature
   X_c is the reactance of the conductor as given for the unarmoured or non-metallic sheathed case in section A.4.3
   $X_m = 0.04355$ mΩ/m at 50 Hz
   F   is the coefficient of coupling between armour and conductor, given by

$$F = \frac{H^2}{R_a{}^2 + H^2}$$

   where:

   $H = X_a - X_m/3$
   $X_a = 0.0628 \log_e(2 \, s/d_a)$ mΩ/m at 50 Hz
   s   is the axial separation of the cables (mm)
   d_a is the mean diameter of the sheath or armour (mm)

A.5 Earth faults

Reactance of line and protective conductors:

i where the protective conductor is provided by a core in a multicore cable
ii where the protective conductor is provided by a separate conductor (but not by the steel-wire armour of a multicore cable or a cable enclosure; see section 6.3 of this Guidance Note).

The reactance of each conductor:

$$X_1 \text{ and } X_p = 0.0628 \log_e(2\ s/a.d) \quad \text{m}\Omega/\text{m at 50 Hz}$$

where s is the axial spacing between the line and protective conductors (mm).

There may be different values of a and d for each conductor, and values of a can be taken from section A.2.1.

Where the protective conductor is provided by a metallic sheath or by the non-magnetic armour of one cable (single-point bonding):

$$R_1 = R_c$$

$$R_p = R_a$$

$$X_1 = 0.0628 \log_e(2\ s/a.d) \quad \text{m}\Omega/\text{m at 50 Hz}$$

$$X_p = 0$$

Where the armour of single-core cables is bonded at both ends of the run, refer to section 6.3.2 of this Guidance Note.

It is assumed that there is no ferromagnetic material between or near to the conductors.

The increase in impedance due to ferromagnetic material close to the conductors depends on the circumstances but may be of the order of 0.03 mΩ/m. It can be kept to a minimum by installing the protective conductor as close as possible to the line conductor (see Booth H.C., Hutchings E.E. and Whitehead S., 'Current Ratings and Impedance of Cables in Buildings and Ships', *Journal of the Institution of Electrical Engineers*, Vol. 83, No. 502, October 1938).

Calculation of k for other temperatures B

B.1 Adiabatic equation

434.5.2
543.1.3

The general formula for k for given initial and final temperatures is:

$$k = K \left(\log_e \frac{\theta_1 + \beta}{\theta_0 + \beta} \right)^{1/2} \qquad (\text{As}^{1/2}/\text{mm}^2)$$

where:

- θ_1 is the conductor final temperature (°C)
- θ_0 is the conductor initial temperature (°C)
- β is the reciprocal of the temperature coefficient of resistance for the conductor material at 0 °C (°C)
- K is a constant for the conductor material (see Table B.1).

Material	β (°C)	K
Copper	234.5	226
Aluminium	228	148
Lead	230	41
Steel	202	78

▼ **Table B.1**
Coefficients β and K for calculation of k

It should be borne in mind that the choice of an alternative value for initial or final temperature in the adiabatic equation should also be taken into account in the calculation of the average conductor temperature for which fault current is determined.

For further information on thermally permissible short-circuit currents and heating effects refer to BS 7454 (IEC 949).

Index

The IEE prepares regulations for the safety of electrical installations for buildings, the *IEE Wiring Regulations* (BS 7671 *Requirements for Electrical Installations*), which have now become the standard for the UK and many other countries. It also recommends, internationally, the requirements for ships and offshore installations. The IEE provides guidance on the application of the installation regulations through publications focused on the various activities from design of the installation through to final test and then maintenance. This includes a series of eight Guidance Notes, two Codes of Practice and Model Forms for use in Wiring Installations.

Requirements for Electrical Installations BS 7671:2008 (IEE Wiring Regulations, 17th Edition)
Order book PWR1700B Paperback 2008
ISBN: 978-0-86341-844-0 **£75**

On-Site Guide (BS 7671:2008 17th Edition)
Order book PWGO170B 188pp Paperback 2008
ISBN: 978-0-86341-854-9 **£22**

Wiring Matters Magazine **FREE**
If you wish to receive a FREE copy or advertise in Wiring Matters please visit
www.theiet.org/wm

IEE Guidance Notes

A series of Guidance Notes has been issued, each of which enlarges upon and amplifies the particular requirements of a part of the IEE Wiring Regulations.

Guidance Note 1: Selection & Erection of Equipment, 5th Edition
Order book PWG1170B 216pp Paperback 2009
ISBN: 978-0-86341-855-6 **£30**

Guidance Note 2: Isolation & Switching, 5th Edition
Order book PWG2170B 74pp Paperback 2009
ISBN: 978-0-86341-856-3 **£25**

Guidance Note 3: Inspection & Testing, 5th Edition
Order book PWG3170B 128pp Paperback 2008
ISBN: 978-0-86341-857-0 **£25**

Guidance Note 4: Protection Against Fire, 5th Edition
Order book PWG4170B 104pp Paperback 2009
ISBN: 978-0-86341-858-7 **£25**

Guidance Note 5: Protection Against Electric Shock, 5th Edition
Order book PWG5170B 144pp Paperback 2009
ISBN: 978-0-86341-859-4 **£25**

Guidance Note 6: Protection Against Overcurrent, 5th Edition
Order book PWG6170B 104pp Paperback 2009
ISBN: 978-0-86341-860-0 **£25**

Guidance Note 7: Special Locations, 3rd Edition
Order book PWG7170B 152pp Paperback 2009
ISBN: 978-0-86341-861-7 **£25**

Guidance Note 8: Earthing & Bonding, 1st Edition
Order book PWRG0241 168pp Paperback 2007
ISBN: 978-0-86341-616-3 **£25**

continues overleaf ▶

Other guidance publications

Commentary on IEE Wiring Regulations (17th Edition, BS 7671:2008)
Order book PWR08640
c.432pp Hardback 2009
ISBN: 978-0-86341-966-9 **£65**

Electrical Maintenance, 2nd Edition
Order book PWR05100
228pp Paperback 2006
ISBN: 978-0-86341-563-0 **£40**

Code of Practice for In-service Inspection and Testing of Electrical Equipment, 3rd Edition
Order book PWR08630
152pp Paperback 2007
ISBN: 978-0-86341-833-4 **£40**

Electrical Craft Principles, Volume 1, 5th Edition
Order book PBNS0330
344pp Paperback 2009
ISBN: 978-0-86341-932-4 **£25**

Electrical Craft Principles, Volume 2, 5th Edition
Order book PBNS0340
432pp Paperback 2009
ISBN: 978-0-86341-933-1 **£25**

Electrician's Guide to the Building Regulations, 2nd Edition
Order book PWGP170B
234pp Paperback 2008
ISBN: 978-0-86341-862-4 **£22**

Electrical Installation Design Guide: Calculations for Electricians and Designers
Order book PWR05030
186pp Paperback 2008
ISBN: 978-0-86341-550-0 **£22**

Electrician's Guide to Emergency Lighting
Order book PWR05020
88pp Paperback 2009
ISBN: 978-0-86341-551-7 **£22**

Electrical training courses

We offer a comprehensive range of technical training at many levels, serving your training and career development requirements as and when they arise.

Courses range from Electrical Basics to Qualifying City & Guilds or EAL awards.

Train to the 17th Edition BS 7671:2008
▶ Update from 16th to 17th Edition
▶ Understand the changes
▶ New qualifying awards C&G/EAL
▶ Meet industry standards

Qualifying Courses
▶ Certificate of Competence Management of Electrical Equipment Maintenance (PAT) – 1 day
▶ Certificate of Competence for the Inspection and Testing of Electrical Equipment (PAT) – 1 day
▶ Certificate in the Requirements for Electrical Installations – 3 days
▶ Upgrade from 16th Edition achieved since 2001 – 1 day
▶ Certificate in Fundamental Inspection, Testing and Internal Verification – 3 days
▶ Certificate in Inspection, Testing and Certification of Electrical Installations – 3 days

Other 17th Edition Courses
▶ Earthing & Bonding – For designers and electrical contractors who require a good working knowledge of the E & B arrangements as required by BS 7671:2008
▶ 17th Edition Design – BS 7671 and the principles associated with the design of electrical installations

To view all our current courses and book online, visit **www.theiet.org/coursesbr**

To discuss your training requirements and for on-site group training, please speak to one of our advisors on +44 (0)1438 767289

Order Form

Details

Name:

Job Title:

Company/Institution:

Address:

Postcode: Country:

Tel: Fax:

Email:

Membership No (if Institution member):

Payment methods

☐ By **cheque** made payable to The Institution of Engineering and Technology

☐ By **credit/debit card:**

☐ Visa   ☐ Mastercard   ☐ American Express   ☐ Maestro Issue No:_____

Valid from: ☐☐ ☐☐   Expiry Date: ☐☐ ☐☐   Card Security Code: ☐☐☐☐
(3 or 4 digits on reverse of card)

Card No: ☐☐☐☐ ☐☐☐☐ ☐☐☐☐ ☐☐☐☐

Signature_____ Date _____
(Orders not valid unless signed)

Cardholder Name:

Cardholder Address:

Town: Postcode:

Country:

☐ By official **company purchase order** (please attach copy)
EU VAT number:_____

Ordering information

Quantity	Book No.	Title/Author	Price (£)
		Subtotal	
		- Member discount**	
		+ Postage /Handling*	
		+ VAT (if applicable)	
		Total	

Membership

Passionate about engineering? Committed to your career?

Do you want to join an organisation that is inspiring, insightful and innovative?

One of the most highly recognised knowledge sharing networks in the world, membership to the Institution of Engineering and Technology (IET) is for engineers and technologists working or studying in an increasingly multidisciplined, digital and global environment.

Joining the IET and having access to tailored products and services will become invaluable for your career and can be your first step towards professional qualifications.

You could take advantage of …

▶ 18 issues per year of the industry's leading publication, *E&T* magazine.

▶ Professional development and career support services to help gain professional registration.

▶ Discounted rates on dedicated training courses, seminars and events covering a wide range of subjects and skills.

▶ Watch live IET.tv event footage at your desktop via the internet, ask the speaker questions during live streaming and feel part of the audience without physically being there.

▶ Access to over 100 local networks around the world.

▶ Meet like-minded professionals through our array of specialist online communities.

▶ Instant online access to over 70,000 books, 3,000 periodicals and full-text collections of electronic articles – wherever you are in the world.

▶ Discounted rates on IET books and technical proceedings.

Join online today at www.theiet.org/join or contact our membership and customer service centre on +44 (0)1438 765678

Professional Registration

What type of registration is for you?

Engineering Technicians (EngTech) are involved in applying proven techniques and procedures to the solution of practical engineering problems. You will carry supervisory or technical responsibility, and are competent to exercise creative aptitudes and skills within defined fields of technology. Engineering Technicians also contribute to the design, development, manufacture, commissioning, operation or maintenance of products, equipment, processes or services.

Incorporated Engineers (IEng) maintain and manage applications of current and developing technology, and may undertake engineering design, development, manufacture, construction and operation. Incorporated Engineers are engaged in technical and commercial management and possess effective interpersonal skills.

Chartered Engineers (CEng) develop appropriate solutions to engineering problems, using new or existing technologies, through innovation, creativity and change. They might develop and apply new technologies, promote advanced designs and design methods, introduce new and more efficient production techniques, marketing and construction concepts, pioneer new engineering services and management methods. Chartered Engineers are engaged in technical and commercial leadership and possess interpersonal skills.

For further information on Professional Registration (EngTech/IEng/CEng), tel: +44 (0)1438 765673 or email: membership@theiet.org

Notes

Notes

Notes

Notes